"What a wonderfully affirming and informative addition the library of every teacher of secondary mathematics! *There Is No One Way to Teach Math* attends to all the right topics as it practically illuminates the importance of discussion, collaboration, variety, tools, and bridging the unhelpful either-ors with common-sense balance".

**Steven Leinwand**, *American Institutes for Research, USA*

"*There Is No One Way to Teach Math* is humane, validating, and wise. It reads like a conversation with the master teacher down the hall: grounded, realistic, and refreshingly honest. I look forward to assigning this nutrient-rich book as required reading in my course for in-service and pre-service teachers".

**Amanda Cangelosi**, *Utah State University, USA*

# There Is No One Way to Teach Math

A collaboration between a seasoned math teacher and a research mathematician, this resource offers balanced instructional ideas based on student intellectual engagement and skilled teacher leadership. It is solidly grounded in many areas of classroom practice, but rather than serving as a prescriptive how-to manual, the authors invite reflection and discussion across classrooms and math departments, much in the way you would share ideas in the teachers' lounge or across the table at a conference. Chapters offer practical suggestions and concrete examples to teachers of grades 6–12 on just about every aspect of the job: manipulatives, technology, lesson planning, group work, classroom discussion, and more. In opposition to the idea of a "one-size-fits-all" curriculum, the authors explain how to integrate teaching techniques: formal and informal, student-centered and teacher-led, experiential and rigorous. Chapters also include vignettes, as well as many links to curricular materials. Ideal for math educators of grades 6–12, this book is both comprehensive in its strategies and sensitive to the complexities of teaching. For these reasons, math departments, coaches, teacher leaders, and faculty at other levels can also easily reference its content where relevant. This book offers multiple entry points for teachers and departments to discuss and enhance their practice, making it essential reading for any math classroom or professional development opportunity.

**Henri Picciotto** is a retired teacher, having spent 42 years as a K–12 math teacher, mostly in grades 9–12. As a consultant, he has worked with more than 50 schools and offered hundreds of workshops.

**Robin Pemantle** is a professor of mathematics at the University of Pennsylvania. Previously, he has taught math enrichment to grades 5–8 and was a Lilly Teaching Fellow.

# *Other Eye On Education Books Available from Routledge*

(www.routledge.com/eyeoneducation)

**Reaching & Teaching Neurodivergent Learners in STEM: Strategies for Embracing Uniquely Talented Problem Solvers**
*Jodi Asbell-Clarke*

**Introducing Nonroutine Math Problems to Secondary Learners: 60+ Engaging Examples and Strategies to Improve Problem-Solving Skills**
*Robert London*

**Exploring Math with Technology**
*Allison W. McCulloch and Jennifer N. Lovett*

**Mathematics Teaching on Target**
*Alan Schoenfeld, Heather Fink, Alyssa Sayavedra, Anna Weltman, Sandra Zuniga-Ruiz*

**A Sensory Approach to K-6 STEAM Integration**
*Kerry Holmes, Jerilou J. Moore, Stacy V. Holmes*

**STEAM Teaching and Learning Through the Arts and Design**
*Debrah Sickler-Voigt*

**Math Problem Solving in Action: Getting Students to Love Word Problems, K-2**
*Dr. Nicki Newton*

**Math Problem Solving in Action: Getting Students to Love Word Problems, 3-5**
*Dr. Nicki Newton*

**Guided Math Lessons in Fifth Grade: Getting Started**
*Dr. Nicki Newton*

**Day-by-Day Math Thinking Routines in Fifth Grade: 40 Weeks of Quick Prompts and Activities**
*Dr. Nicki Newton*

**Guided Math in Action: Building Each Student's Mathematical Proficiency with Small-Group Instruction, Second Edition**
*Dr. Nicki Newton*

# There Is No One Way to Teach Math

*Actionable Ideas for Grades 6–12*

Henri Picciotto and Robin Pemantle

Taylor & Francis Group

NEW YORK AND LONDON

Cover image: Briana Loewinsohn

First published 2025
by Routledge
605 Third Avenue, New York, NY 10158

and by Routledge
4 Park Square, Milton Park, Abingdon, Oxon, OX14 4RN

*Routledge is an imprint of the Taylor & Francis Group, an informa business*

© 2025 Henri Picciotto and Robin Pemantle

The right of Henri Picciotto and Robin Pemantle to be identified as authors of this work has been asserted in accordance with sections 77 and 78 of the Copyright, Designs and Patents Act 1988.

All rights reserved. No part of this book may be reprinted or reproduced or utilised in any form or by any electronic, mechanical, or other means, now known or hereafter invented, including photocopying and recording, or in any information storage or retrieval system, without permission in writing from the publishers.

*Trademark notice:* Product or corporate names may be trademarks or registered trademarks, and are used only for identification and explanation without intent to infringe.

ISBN: 978-1-032-75406-2 (hbk)
ISBN: 978-1-032-75933-3 (pbk)
ISBN: 978-1-003-47385-5 (ebk)

DOI: 10.4324/9781003473855

Typeset in Palatino
By KnowledgeWorks Global Ltd.

# Contents

*Preface* .................................................... xi
*Acknowledgments* ............................................ xiii

**Introduction** .............................................. 1
What Is and Is Not in This Book  1
Our Assumptions  2
How to Use This Book  4
Notes  5

## PART I  Pedagogical Principles .............................. 7

### 1  Philosophical Framework ................................. 9
Embracing Contraries  9
Compliance  13
Teaching for Understanding  14
Number Sense and Other Senses  17
Note-Taking and Institutionalization  18
All of the Above!  21
Discussion Questions  21
Notes  21

### 2  Problem Solving ........................................ 23
Two Sample Problems  24
Placing Problems in the Curriculum  29
Inquiry  31
Rich Activities  34
Creating Problems  37
Using Puzzles  42
Problem Solving at the Core  46
Discussion Questions  47
Notes  47

### 3  Different Modes ........................................ 49
Classroom Basics  50
Openers  52
General Routines  55
Kinesthetic Activities  56

◆ vii

Games  57
Diversifying One's Portfolio  58
Discussion Questions  58
Notes  59

## PART II   Classroom Practice .............................. 61

### 4  Cooperative Learning ...................................... 63
Group Work: Why  64
Group Work: How  65
Common Obstacles  69
Dead Ends  72
Individual Growth  75
Discussion Questions  76
Notes  76

### 5  Learning Tools .......................................... 77
Why Tools?  77
Which Tools?  80
Conceptual Tools  82
Multiple Representations  93
Seeing Is Believing?  95
Active Involvement  96
Discussion Questions  97
Notes  97

### 6  Manipulatives .......................................... 99
Models  99
Geometric Puzzles  103
Even More Manipulatives  110
From the Concrete to the Abstract  114
Discussion Questions  115
Notes  115

### 7  Computation Engines ..................................... 117
A Brief History  117
Graphing Technology  121
Interactive Geometry  127
Computer Algebra Systems  130
Virtual Manipulatives  132
Tools as Curriculum  133
Robot Teachers?  137
Discussion Questions  137
Notes  138

## 8 Leading Discussions . . . . . . . . . . . . . . . . . . . . . . . . . . . . . . . . . . . 141
Project SEED  142
Every Minute Counts  145
Sequencing Discussions  147
Staging  151
Encouraging Without Babying  152
Handling Wrong Answers  154
Listening  155
Questioning  157
Where We Are  162
Discussion Questions  162
Notes  162

## PART III   The Big Picture . . . . . . . . . . . . . . . . . . . . . . . . . . . 165

## 9 Extending Exposure. . . . . . . . . . . . . . . . . . . . . . . . . . . . . . . . . . 167
Reaching the Full Range  167
Guiding Principles  169
Lagging Homework  170
Review  173
More on Review  174
Discussion Questions  175
Note  175

## 10 Planning. . . . . . . . . . . . . . . . . . . . . . . . . . . . . . . . . . . . . . . . . . . 177
Pruning  177
A Framework for Planning  178
Mapping Out a Course  183
Spiraling  185
Pursuing Two Units Simultaneously  186
Learning from Experience  187
Discussion Questions  188
Notes  188

## 11 Assessment . . . . . . . . . . . . . . . . . . . . . . . . . . . . . . . . . . . . . . . . 189
The Future of Grades  190
Tests and Quizzes  190
Retakes Versus Test Corrections Versus Neither  192
Other Assessments  195
Yet More Options  196
Assessment and Extending Exposure  197
Discussion Questions  198
Notes  198

**12 Making Change** .............................................. 199
   Course and Teacher Evaluations  199
   Collaboration  201
   Research  203
   Fads Versus Eclecticism  204
   Setting Goals  206
   Discussion Questions  207
   Notes  207

   *Appendix: Special Note to Young Teachers* ........................ 209
   *Index* .......................................................... 213

# Preface

Henri has retired from the classroom after 42 years as a K-12 math teacher, and now works as a consultant, sharing ideas and materials online and in workshops. Robin is a research mathematician at the University of Pennsylvania, with a strong interest in education. We met in the 1970s when Robin was in middle school, and Henri introduced him to some "math enrichment" topics. Many years later we reconnected, and upon seeing both agreements and disagreements in our views on teaching, we decided to try to write a book together.

Our first attempt yielded an unwieldy tome which we hoped would be encyclopedic. This didn't work out as well as we had hoped, and we decided to split it in two. One book (*Beyond the Math Wars*) would be addressed to the general public, and focus on curriculum and policy. That book is currently a work in progress. The other book (this one) would be geared to math teachers, and focus on pedagogy. If you're a math teacher, we hope you will read both.

In our view, there is no single best way to teach math (or any subject). There is too much variability among students, classes, and schools. And likewise, teachers have different backgrounds and experiences, different personalities, and different strengths and weaknesses. Given all this, we have to accept that no strategy or technique works in all situations.

And yet, all of us have learned a lot from colleagues – at school, at conferences, online. Sometimes, what we learn becomes useful to us, and more importantly, to our students. It is in that spirit that we offer the ideas in this book: consider it a source of possibly useful ideas to incorporate into your own teaching practice and in the discussions with your colleagues – ideas that emerged in extended conversations between a math teacher and a mathematician, based on our experiences, values, and readings.

This last point is the other motivation for this book. Too often, teachers and mathematicians have engaged in counterproductive polemics. We see this book as the result of a more constructive, collaborative practice. We hope to convince you to turn away from the "math wars" that have been endemic in this culture and join us in practicing an eclectic, non-dogmatic, and balanced approach to the art of teaching.

# Acknowledgments

Cover illustration by Briana Loewinsohn.

Many thanks to Amanda Cangelosi, Anita Wah, Anna Weltman, Dan Bennett, Frances Kandl, James Propp, Jennifer Szydlik, Liz Caffrey, Marilyn Burns, Meghan Lee, Michael Pershan, Oscar Pemantle, Parisa Safa, Rachel Chou, Richard Lautze, Scott Farrand, Scott Nelson, Steve Leinwand, Susan Moon, and Vic Ferdinand.

And to these institutions: Black Pine Circle Day School, Ohio State University, the University of Washington's Whiteley Center, the Urban School of San Francisco, and the University of Wisconsin.

# Introduction

**What Is and Is Not in This Book**

The curriculum is *what* we teach. There are some disagreements in our profession about that, but they are minimal if compared with disagreements about pedagogy. Pedagogy is *how* we teach. There is a growing consensus among our professional organizations which puts student thinking, exploration, and discussion at the heart of this enterprise. However how to achieve this is not obvious, and not everyone even agrees that this should be our goal.

In spite of what is said at conferences, and written in journal articles, we suspect that many math classes in the US still follow a pattern that in our view is not particularly effective. It goes something like this: the teacher introduces an idea or technique. Students are expected to pay close attention, practice in silence, and make sure they remember what the teacher said. That widespread approach is problematic. Without some understanding, it is not easy to remember the concepts and techniques – there is just too much to remember. Understanding is the glue that connects the ideas and makes them stick. On the other hand, a common overreaction to the traditional exclusive reliance on teacher explanations can lead to an equally counterproductive abdication of teacher responsibility.

In this book we share our experiences and beliefs on how to engage students intellectually – the basic prerequisite for lasting understanding, and on how teacher leadership is essential to reap the benefits of that approach. Most of our examples will be drawn from the teaching of middle school and high school math, but much of what we write will apply to elementary school and college, as we have had experience at all those levels.

Inevitably, our thoughts about curriculum and educational policy will appear here and there throughout the book, but that is not what the book is mostly about. Our emphasis will be overwhelmingly on pedagogical questions. We will not say much about topics on which our expertise is limited, such as classroom management and standardized tests, but we will discuss just about every other aspect of math teaching.

While we will cite various educators who influenced us, we will not be shy about referencing Henri's website and his blog posts.[1] In fact, much of the material in this book originated there.

## Our Assumptions

Before we actually get started, we should share some of our assumptions. Those are spelled out in much greater detail in *Beyond the Math Wars*.

### The Math Wars

In this book, we do not go into historical context. However, we do need to explain the phrase "Math Wars". For decades there has been an intellectual division about the best ways to teach math and, to some degree, about the proper aims of mathematical education. One side stood for basic skills, knowledge of procedures, measurable abilities, and polished performance. The other side stood for conceptual understanding, problem solving, "mathematician's math", and the discovery process. Sometimes one of these sides was in ascendancy, sometimes the other, as large-scale reforms rocked math education from one extreme to the other. Some reforms big enough to have names were: the New Math (circa 1960), Back to Basics (circa 1982), reforms driven by the 1989 and 2000 National Council of Teachers of Mathematics (NCTM) Standards, No Child Left Behind (2002), and the Common Core State Standards (2010).

The two sides tend to align with conservative and liberal political ideologies. This is quite unfortunate because this alignment probably serves to increase the polarization felt by math education stakeholders, and exacerbates the tendency to pursue one-sided theories of math education. This only got worse when the polarization of math educational theory met up with the explicitly political culture wars concerning values being conveyed through public school systems. Although these divisions affect math only tangentially, they sometimes come to be topics of urgent and hostile debate. Specific hot-button issues in math education include testing, technology, tracking, and traditional versus new curricular paths.

Our view is that good math teaching involves all the aspects cherished by both sides of the debate and that you cannot survive on some of these without the others. Why choose between basic skills and conceptual understanding when either one on its own is useless? Chapter 1 sets up the idea of embracing contraries and the book proceeds from there, leaving the Math Wars behind.

## Habits of Mind

Our job as math educators is to teach math, naturally. But this cannot be done effectively if we do not also teach some necessary habits of mind. Habits of mind have been an ongoing concern of math educators for a long time, and they have been described in various ways, the most prominent being NCTM's "Process Standards", and the Common Core "Standards for Mathematical Practice". We like those formulations, and others,[2] but in *Beyond the Math Wars*, we detail a version of this that we call the meta-curriculum. Here is a brief description of some of its components.

*Concentration* is the ability to hold a lot of information and relationships in mind, to avoid distraction, and to follow a thought to its end. For younger students, better concentration means being able to pay attention to a problem and remember it. As students progress, better concentration will help them with abstraction and with the ability to make connections to other knowledge.

*Abstraction* is the ability to see something as an instance of a general schema and to understand the logical consequences of its being such an instance. In some sense, how far a student can go in math is governed by the level of abstraction that the student has learned to deal with. Addition is an abstraction. Subtraction (what would I have to add to 2 to get 5?) adds a layer. Algebra, where variables represent numbers, adds another layer. Functions add another level, and calculus yet another level.

*Initiative* means students taking responsibility for their own learning. If you don't know something, do you try and think it through, come up with questions, try something even if it might not work, or do you wait for it to be explained? Waiting leads to less learning because there is no engagement, and because the explanation itself might require explanation.

*Ownership* of knowledge is another piece of the meta-curriculum. A student who owns the knowledge feels that it makes sense, that they could perhaps have figured it out on their own, that they could explain it to someone else, and that they have enough confidence to still believe it even if the teacher were to rescind the lesson and say they taught something wrong.

*Articulation* is the ability to formulate mathematical ideas. It is a communication skill because you can't discuss or argue mathematics until you can articulate it but it is also an internal skill: you haven't learned something all the way until you can articulate it to yourself. *Precision* is a sub-skill of articulation: using mathematical language to say something unambiguous. *Argumentation* is the articulation of justification. At its most formal, argumentation is proof, but there are many levels of argumentation before this. The first is to express why you believe something. A more compelling argument may be needed to convince a friendly skeptic. Higher levels of argument address logical structure, examples, or counterfactuals.

Central to supporting the meta-curriculum is the practice of *problem solving,* to which we dedicate Chapter 2, and in fact much of this book. This practice, long-championed by NCTM, provides a context for math learning by triggering student intellectual engagement. In the end, habits of mind are more important than any one piece of curriculum, as they are helpful not only in math class, but also in other disciplines, and in fact outside of school.

**Heterogeneous Classes**

We are well aware of the social impact of tracking math classes into (e.g.) "honors" vs. "regular" classes. This widespread practice often results in a rigid hierarchy within our schools and reinforces socio-economic, gender, and racial inequalities. On the other hand, we also know that teaching untracked classes is extraordinarily challenging, and requires teachers who have a strong understanding of both subject matter and pedagogy.

Many of the ideas we share in this book are aimed at supporting the teaching of heterogeneous classes. Of course, even tracked classes are heterogeneous, so we are confident that our suggestions will be broadly useful.

## How to Use This Book

Because the book covers the everyday work of math instruction from so many angles, it can be used in one of two ways:

- If you are trying to figure out an overall approach to your practice, we present a coherent set of guidelines to support student inquiry and teacher leadership.
- If you seek specific techniques to implement as a next step in your professional development, we offer a wealth of ideas on almost every aspect of the job – peruse the table of contents to find a specific area to focus on.

However, it is difficult to grow as a teacher if one works in complete isolation. This is why we end each chapter with *discussion questions*. Given the complex nature of our chosen profession, everything anyone says about it deserves scrutiny. Do you agree with the ideas we shared in the chapter? Do they apply to your school? Will they work for you and your students? Those questions can be taken up by math departments, in a buddy system among two or more teachers, in a preservice course on math pedagogy, or in a professional development workshop.

**Notes**

1. www.MathEducation.page and blog.MathEducation.page
2. See A. Cuoco, E. P. Goldenberg, & J. Mark. Habits of mind: An organizing principle for mathematics curricula. *Journal of Mathematical Behavior,* 15:375–402, 1996. And also A. Pickford. withoutgeometry.com/2010/09/habits-of-mind.html

# Part I
# Pedagogical Principles

# 1
# Philosophical Framework

An ironic vignette from *The Mathematics Teacher*[1]:

> I learned to ride a bike at age six. The experiences leading to my first solo bike ride are still vivid memories. First, my father gave a wonderful explanation of bike riding. The mechanics of leg motion required to pedal were explained. Hand placement on the handlebars, and the nuances of steering were described with clarity. Braking, maintaining balance, and every other conceivable aspect of bike riding were laid before me. Next, my brother modeled bike riding. Down the block he rode: pedaling, steering, maintaining balance. He managed a 180-degree turn, returned, braked, and came to a perfect stop. Having gained a conceptual understanding of bike riding and having observed successful bike riding, I was able to ride the bike by myself at first attempt.

Would that it were so easy! You cannot learn to ride a bicycle without actually riding it. You cannot learn math without actually doing math.

This applies as much to teaching as it does to learning. No amount of reading about classroom techniques can possibly make you good at them. This is the job of teacher preparation programs, of continuing professional development, of individual mentors and colleagues, and of classroom experience. We hope this book will support all that!

In this chapter, we present an overview of our philosophy. After rejecting simplistic binaries, we state and disentangle our central goal: to teach for understanding. We end with an acknowledgment that while understanding is fundamental, it is not enough: students must be brought into the institution of mathematics, which has its own culture and conventions.

## Embracing Contraries

During the Math Wars of the 1990s, an absurd debate developed. A minority of reformers, overreacting to the mindless memorization at the center of much traditional math education, deliberately de-emphasized skill development. A backlash ensued, charging that the entire reform movement was abandoning skill-building and that understanding could only come after skills had been mastered. In fact, both sides are wrong: understanding cannot be divorced from the acquisition of skills. Without understanding, it is hard to develop an interest in the skills or to retain them; without skills, understanding is out of reach. Good teaching requires weaving those two strands together.

The skills-versus-concepts vendetta is curricular, but the same rhythm can be seen in pedagogical fads. Consider their life cycle: researchers uncover an important idea about how children learn. Word spreads. Administrators want to bring methods based on these results into their schools and districts. Common sense says to sprinkle these into the mix, but a combination of wishful thinking and desperation says to go the whole hog. Throw out the old ways! Some consultants tap into this phenomenon and become the carriers of the new gospel. The research results, even if they were valid in the setting in which they were discovered, become vastly overgeneralized. The story does not end well. The irony is that when we botch the implementation, we lose the wisdom behind the idea. There is innovation, sometimes brilliance, in many education fads. Our job is to be eclectic, to appreciate the efforts behind these, and to separate the useful from the ridiculous.

In 1983, Peter Elbow, a professor at Stony Brook University, wrote a profound article about teaching.[2] The article helped us clarify our thinking about math education even though Elbow is a teacher of English. In his second paragraph, Elbow says,

> ... the two conflicting mentalities needed for good teaching stem from the two conflicting obligations inherent in the job: we have an obligation to students but we also have an obligation to knowledge and society. Surely we are incomplete as teachers if we are committed only to what we are teaching but not to our students, or only to our students but not to what we are teaching, or half-hearted in our commitment to both.

To fully grasp the significance of this, you have to see how often these two obligations directly oppose each other. For example, our commitment to mathematics requires us to dissect a student's argument with a relentlessly critical eye, tightening the logic and clarifying the language. At the same time, our commitment to the student requires us to be encouraging and supportive, to look for whatever germs of good thinking are there, even if the argument is flawed, and to find ways to build on those. These are nearly opposite, yet we need to be totally committed to both.

A given teacher often leans toward one of these yin and yang opposites, because of personality, philosophical bent, or cultural milieu. We suggest that effective teachers constantly broaden their repertoire, and learn to make choices that may not come naturally. The first step is the most important: identifying and embracing contraries is what allows a teacher to navigate between the two extremes.

Henri made a rather complete list of such contraries.[3] Here, we'll discuss a few examples: "soft" versus "hard" teaching, forward motion versus review in course pacing, new versus familiar lessons in planning, and direct instruction versus discovery. This list is far from comprehensive, but it will give you a sense of how to approach contraries. We return to all these contradictions later in the book.

### Soft Versus Hard Teaching

To finish Elbow's example, so-called soft teachers prioritize their commitment to the students; so-called hard teachers prioritize their commitment to the discipline. Good teaching requires not a compromise between the two but an ability to do both at different times. Elbow suggests that at the start of a course, you should be hard and make clear your criteria for grading. After that, you should be soft and 100% on the student's side to help them achieve the goals you laid out. And finally, at the end of the course, you should be hard again, and give them the grade they deserve. Elbow's ideas are a useful construct, though somewhat oversimplified. The work of a teacher is to be both encouraging and honest in offering feedback. This is a constant back-and-forth struggle, requiring constant navigation along this axis – sometimes emphasizing one commitment, sometimes the other. (We discuss assessment in Chapter 11.)

### Course Pacing

On the one hand, course pacing requires forward motion. Without that, we lose the motivation of our strongest classroom allies, the ones who love our subject and can't get enough of it. Actually, most students' morale will suffer if the class remains mired in old material. But it is also our responsibility to build in an adequate amount of review, which most students absolutely need in order to achieve mastery. It seems kind to go at a snail's pace, but in fact, it is not. A breakneck pace would seem to lead to excellence, but in fact it does not. We must embrace both constant forward motion and eternal review, and avoid always choosing one over the other. In Chapters 9 and 10, we share ideas on how this might be carried out to achieve a win on both counts.

You probably noticed that this pair of opposites is a different manifestation of obligation to the discipline (forward motion) versus obligation to the students (review). Another manifestation of this same tension is between covering the material versus responding to the realities of the class. Allowing either to always dominate is another form of betrayal.

There is no way to have a one-size-fits-all version of a course, because every class is different, and in fact, even different sections of the same class are different. Good teaching requires skillful navigation between those poles.

**New Lessons Versus Old**

Take lesson planning. Doing a given lesson in a new way is stimulating for us as teachers, and some of that electricity is passed on to the students. On the other hand, the amount of time and energy we have at our disposal is finite, and we cannot and should not constantly reinvent our lessons. If something worked last year, it is not a sin to try it again this year. Once again, we must embrace opposites, and make judgment calls about when to be creative and when to be efficient. As King Solomon (and Pete Seeger) put it, "to everything there is a season".

**Lecture Versus Activity, Direct Instruction Versus Discovery**

The ideological divide on this issue is huge. On one side we hear, "Students must be actively engaged in order for them to internalize the lesson. Lecture is a way to pass information from the teacher to the notebook, without entering the mind of the student". On the other side, we hear, "Students need a clear statement of basic facts. Student-invented methods are inefficient to develop, lack commonality, and even when correct are typically inferior to the professional consensus developed over the centuries". Both sentiments have some truth. Teachers make decisions daily that balance these needs: not only lecture versus activity but also formal versus informal, answer versus process, textual versus hands-on, controlling the conversation versus running with it, et cetera. Much of this book is about achieving a nuanced understanding of these choices.

It is not true that anything students "discover", they will remember. First of all, what we hope they discover may not be what they actually understand. But also, it's not clear to them what is important about their discovery, what is worth remembering, how it connects with other concepts, and so on. (We return to this issue at the end of this chapter.)

On the other hand, merely providing good explanations does not turn out to be any more effective, as students don't necessarily listen to those, and even if they do, they may not understand them. Thus, they are forced to memorize poorly understood techniques and ideas. Because there is so much math to be learned, memorization often only works until the quiz. A few weeks later, it's mostly gone.

As teachers, we have had to learn how to combine student inquiry with teacher guidance. We reject the hardcore "never give a hint" position of people who overestimate inquiry. We also reject the belief that excellent teacher explanations suffice. Effective teaching requires us to navigate between many strategies which appear to be mutually exclusive, but are in fact complementary. Well-chosen problems help students engage with the question at hand, and prime them to listen to and understand teacher explanations. How to choreograph this back-and-forth dance is learned through practice, teacher observation, and collaboration. Much of this book consists of specific suggestions on how to manage this.

Unfortunately, a culture of "I explain it, then you practice in silence" still dominates too many math classes, leaving no room for student intellectual engagement. In other classes, students are left to struggle without any guidance. We hope this book will help move its readers toward a more balanced approach.

## Compliance

So yes, we have to embrace opposites. But there are some widespread practices that in our view serve neither the students, nor the discipline. Those are the practices that prioritize student compliance.

Some are obviously unrelated to learning math. Penalizing students for placing the staple in the wrong place, not putting a box around the final answer, using the supposedly wrong writing implement, or any similar violation is absurd. Sure, some of those sorts of requirements facilitate our lives as teachers, and there's no reason not to explain that to the student. But it is ridiculous to have students lose "points" over them. To state the obvious: compliance is not understanding.

There are some practices, which on the surface appear to be about math, but in fact are about enforcing obedience. Here are some examples:

- *Overemphasis on terminology, and unnecessarily complicated notation.* For example, asking kids to learn the words "minuend" and "subtrahend". (Fortunately, that does not seem to have survived in contemporary curricula.) Or insisting on a different notation for a segment and its measurement.
- *Rejecting correct answers that have not been "simplified"* (fractions, radicals), or are in the wrong format. Students make enough real mistakes – let's not tell them they're wrong when they're right! Example: writing polynomials not in descending order.
- *Memorization without understanding*: if students are asked to memorize some facts or algorithms that mean nothing to them, it's just an exercise in blind obedience.

Yes, there is a place for all those things. Distinguishing "equal" from "congruent" makes sense for polygons but not so much for line segments. Lowest terms fractions and simple radical form can be useful, and they can help communication, but they should not be elevated beyond that. Memorized shortcuts and formulas are worthwhile if they encapsulate understanding, but not if they are a substitute for it. And so on.

In general, when we make pedagogical and curricular choices, we should ask ourselves: does this choice support the math? Does it support my students? Or is it more about controlling the students and enforcing compliance? If we are strong in our commitment to our discipline and to our students, we can dramatically decrease our preoccupation with obedience.

## Teaching for Understanding

The solidity of the foundation determines how high you can build. Fragile fundamentals explain attrition from math in high school and college. Taking enough time to get across foundational understandings sometimes appears to undermine short-term demands for performance. But in fact, it supports the long-term best interest of students. In the very short term, the apparent contradiction is unavoidable, but in the span of just a few years (out of 12 or 13!) building a solid foundation will manifest in the short-term goals as well, and students come out ahead on both.

One obstacle to teaching for understanding is that understanding cannot easily be conferred by explanations. Try to write a sentence explaining what a variable is. Try a paragraph. There is no perfect explanation that will bring about this understanding. A student needs to try using variables in order to understand how they work. In the simplest case, they are missing values to be inferred. They follow rules similar to numbers, if properly generalized and if notation can be agreed upon. Sometimes they vary together, sometimes independently. Sometimes they have a single but unknown value, as in the equation $x + 2 = 5$, while at other times they could take on any value as in the expression $3x - 4$ or the identity $x + x = 2x$.

Another obstacle to teaching for understanding is that understanding is not always valued by students, parents, and administrators, some of whom believe that everything would be so much more straightforward if students could simply be asked to memorize facts and algorithms.

### What Is Understanding?

This is a difficult question, and the true fact that experienced math teachers can "recognize it when they see it" is not a sufficient answer. Here is an attempt at spelling it out. A student who understands a concept can:

- *Explain it*. For example, can they give a reason why $2(x + 3) = 2x + 6$? Responding "it's the distributive rule" is evidence that the student knows the name of the rule, but a better explanation might include numerical examples, or a figure using the rectangle model, or a manipulative or visual representation. We should routinely ask students to explain answers, verbally or in writing, even though many don't enjoy doing that. It is a way for us to gauge their understanding, and thus improve our teaching, and more importantly, it is a way for them to go deeper and guarantee the ideas stick.

- *Reverse processes* associated with it. For example, a student does not fully understand the distributive law if they cannot factor anything. More examples: can they create an equation whose multi-step solution is 4? Can they figure out an equation when given its graph? And so on. Reversibility is a test of understanding, a way to improve understanding, and in some cases an alternate path to understanding.
- Flexibly *use alternative approaches*. For example, for equation solving, in addition to the usual "do the same thing to both sides" for solving linear equations, students should be able to use trial and error, graphs, tables, and technology. If they have this flexibility, they can decide on the best approach to solve a given equation, and moreover, they will have a better understanding of what equation solving actually is.
- *Navigate among multiple representations* of it. Famously, functions can be represented symbolically, or in tables, or in graphs. Making the connections between these three is a hallmark of understanding. Multiple representations, on the one hand, offer different entry points that emphasize different aspects of functions, but making the connections between the representations is part and parcel of a deeper understanding.[4]
- *Transfer* it to different contexts. For example, ideas about equivalent fractions are relevant in many contexts, such as similar figures and direct variation. Or, the Pythagorean theorem can be used to find the distance between two points, given their coordinates. If a student can only handle a concept in the form it was originally presented in class or in the textbook, then surely no one would claim they fully understand it.
- *Know when it does not apply*. When faced with an unfamiliar problem, students will tend to reach for familiar concepts, such as linear functions and proportional relationships. Sometimes, this makes sense, of course, but students need to be able to recognize situations where a given concept does not apply. Including "non-examples" when teaching a concept can help with this. For example, what are operations for which the distributive property does *not* apply?

Clearly aiming for all this is a high bar, and it is tempting to just have students memorize some facts and techniques, and then test them to see if they remember those a few weeks later. But what good would that do? It would just add to the vast numbers who got A's and B's in secondary school math, went to college, and now tell us "they're not math people". Yes, teaching for understanding is ambitious, and it must be our goal for all students.

Every once in a while, especially during "math wars", newspapers carry op-eds about math education, arguing that students must master basic skills *before* they can develop conceptual understanding. And moreover, that the road to such mastery is a teacher's explanation followed by repetitive drill. These essays frequently argue that it's like learning to play the piano: you must practice scales before playing real music! When Henri mentioned this to a friend who is a piano teacher, he considered it to be an insult to his profession. He said that obviously these people are not piano teachers! Teaching piano is about music! Yes, students do need exercises, but if that's all you had them do, you'd drive them away. The biggest motivator is the recital, when they play real music, not scales.

Henri's friend is right: the authors of these op-eds are not piano teachers, but they're often not math educators either. Still, it is important to address their ideas, because they reflect a broad cultural consensus among many parents, administrators, students, and teachers. Some proponents of the "skills first" approach equate teaching for understanding to what they call "fuzzy math", a flaky anything-goes sort of teaching, with no specific learning goals, no accountability, just feel-good teachers who allow students to wallow in their ignorance. It behooves those of us who disagree with this caricature to clarify what we mean by understanding. Much of this book is about that.

**Understanding Versus Tricks**

Finally, we should say something about what understanding is *not*. We are in general agreement with *Nix the Tricks*[5] a compendium of how to avoid or at least reduce the use of catchphrases in our teaching. Shortcuts like "FOIL" can obscure the underlying mathematics. They sometimes reflect a cynical attitude: "the students will never understand these concepts, so let's give them an easy-to-remember shortcut".

To be clear, this is not a blanket position against mnemonics. Our concern is with the too-frequent use of tricks as an alternative to understanding. If a student does understand the distributive property, and thus can multiply polynomials with any number of terms, there is nothing wrong with using FOIL in the case of a binomial times a binomial. In fact, it helps to automate the mechanics of algebra, so that the student can focus on their work at a deeper level. Moreover, if a student "knows" only FOIL, our task is not to tell them they're wrong. Instead, we want to make it clear to them where this shortcut came from, when it applies, and when it does not. (One way to do that is to use the rectangle model of multiplication, which we discuss in Chapter 5.)

We ourselves do occasionally use a catchphrase to support thinking. For example, in geometry, we might be heard offering the hint: "When working with circles, you should listen to the radii". It is a good hint, which directs the student's attention to crucial properties of the figure they are trying to make sense of. Because of its substantial mathematical content, this is not a trick that should be nixed.

Overall, our stance is that *formulas, shortcuts, and tricks should encapsulate understanding, not substitute for it.*

## Number Sense and Other Senses

One manifestation of understanding in arithmetic is number sense. This involves many components including especially facility with mental calculation, and a grasp of the relative magnitude of numbers. These combine into skill at estimation, and they are largely built on a solid mastery of place value. Number sense is foundational to further work in mathematics, and in fact in all science and technology.

Instead of spending hundreds of hours trying to make children into a poor substitute for a five-dollar app, we should shift to the development of number sense, mental arithmetic, analysis and discussion of various approaches to calculation, and in secondary school, an introduction to number theory. Much can be learned about numbers in those ways. Accuracy and speed are good things, but they are no longer the priority: understanding should be the goal.

A similar mental shift is necessary in algebra. *Symbol sense* is discussed by Abraham Arcavi.[6] His notion of symbol sense includes behaviors such as appropriately choosing to use symbolic manipulation or not when trying to solve a problem; ease in interpreting symbols; efficiently manipulating algebraic expressions; the selection of the right symbol for a given situation; and so on. For example, someone may know how to factor $n^3 - n$, and get $n(n + 1)(n - 1)$, but not see the implication in the particular case where $n$ is an integer. In that case, looking at the expression makes clear that the difference between a number and its cube is the product of three consecutive integers, and therefore includes a multiple of 3 and at least one even number. Therefore, it must be a multiple of 6. In this example, symbol sense and number sense reinforce each other.

It takes work to develop symbol sense. Many beginners in algebra confuse $2x$, $2 + x$, and $x^2$. This is not merely a linguistic obstacle, but also a conceptual and mathematical one. A popular mistake among high school students is to "distribute the square": $(x + 5)^2 = x^2 + 25$. We are not talking about very complicated manipulations, but simple everyday operations like the one above, or like removing the parentheses in $-(y - 1)$. The reality is that even simple manipulations are difficult to understand for many, if not most students.

Some educators hope that symbol sense will develop organically by working on interesting problems. Alas, this is not true. Working with symbols must be taught explicitly. Students who cannot work with symbols are severely handicapped. If students cannot perform simple algebraic manipulations correctly, they cannot effectively pursue math, science, or statistics. This is true whether their lack of facility with symbols stems from being victims of the old-fashioned Algebra 1, which involved memorizing endless lists of ill-understood moves, or from using a curriculum that de-emphasizes symbolism to the point of near-omission. Just like number sense cannot be made obsolete by calculators, symbol sense is a necessary part of math and science, and cannot be made obsolete by technology. Quite the opposite: effective use of the full power of spreadsheets, for example, does require symbol sense.

Conceptually, number sense and symbol sense involve a substantial common component which we would like to call *operation sense*. For example, take the sequence 5, 8, 11, 14, 17, …

- Number sense should include the ability to recognize repeated addition in this sequence, and its relationship to multiplication. (We are adding 3 repeatedly, and by the time we reach the fifth term, we have added 4 times 3.)
- Symbol sense should include the ability to express and recognize the same thing in a form such as $a + nd$, or in this case $5 + 3n$ and moreover the ability to recognize the relationship between that and the general linear function $y = mx + b$ (here $y = 3x + 5$.)

None of this is possible without a solid grasp of addition and multiplication, their structural relationship, and their uses in various applications. An understanding of operations is fundamental to both number sense and symbol sense. Operations are the joint underlying foundation of both arithmetic and algebra. Other senses can be added to the list, such as visual sense, logical sense, and function sense. The "sense" formulation is an improvement in our view of students and learning. It helps move our thinking away from the view of the student as a programmable machine and toward the more realistic idea that students can think and develop intuitions.

### Note-Taking and Institutionalization

Once students develop some understanding, how do we make sure it sticks? Can note-taking help students retain what they learn in exploratory activities? Some teachers at the middle and high school levels favor a proactive approach to teaching students to take notes. Sometimes this skill is one of the most prominent items on their syllabi. We caution against this in math. Henri writes:

> Some time ago, during a professional development workshop, a participant asked how I teach students to take notes in math class. I explained that I did not think students can simultaneously do math, and take notes. It's really one or the other. The teacher gave me a contemptuous look that made clear she disapproved of my answer. Later in the day, *the very same teacher* asked if I was going to share my slides, because she was having trouble both listening and taking notes.
>
> I have led workshops for teachers hundreds of times. "Will you share your slides?" is definitely a Frequently Asked Question. You, dear reader, may have asked it. You certainly have heard someone ask it. It makes sense! The more worthwhile a presentation is, the more you want to give it your undivided attention, and the more you want to have access to all the slides later.

What should we make of this? Teachers are well educated, with many years of experience as students behind them. They know that taking notes interferes with their own ability to focus. And yet, many of these adults believe that their teenage students should be able to multitask: take down information, and simultaneously engage in challenging intellectual work. That is unrealistic: in that situation it is the clerical task that wins, and the intellectual engagement that suffers. Moreover, note-taking creates a feeling that it is not necessary to engage intellectually since the notes will be available to peruse later. The student's participation becomes mechanical, not mathematical. And for many students, "later" never comes, as interpreting the notes is not necessarily easy or motivating. In our experience, few students go home and engage intellectually with the notes they have taken. If you believe otherwise, you are probably overly optimistic.

This fixation on teaching note-taking is a largely unquestioned part of the landscape of secondary school math education. It may reflect certain beliefs about how math is learned. In our view, you learn math by doing math, not primarily by listening carefully and taking notes. The main part of any math class should consist of students solving problems.

### Institutionalization

And yes, of course, at a certain point it is important to put what is learned into words. That itself is an intellectual challenge for students, which can be taken up during a teacher-led discussion. After many students have spoken, it is essential for the teacher to summarize, using standard terminology and notation. That phase is what French math educators call *institutionalization*: bringing students into the vast international institution of mathematics, which has its own language, notation, and criteria to determine the validity of ideas.

If, in addition to this, the teacher shares slides, saved whiteboard files, reference sheets, or interactive notebook pages, that's all good. But the key is that students should spend most of class time doing math. If note-taking during a lecture is really an important skill, students should learn it in a lecture-based class, not in a math class.

Knowing that the punch line will be spelled out clearly later allows students to keep their focus on problem solving during the main part of the class. It is in that problem-solving part of the lesson that the mental infrastructure is built that makes it possible to understand the final summary when it comes. Moreover, it guarantees that the day's (or week's) curricular goals are met, and that the class can continue to move forward. A skilled teacher can build this wrap-up on a foundation of the intellectual work the students have been doing up to that point – something that a textbook or a prerecorded video cannot do.

Remember embracing contraries? Student discovery has many benefits, but alone, it guarantees neither understanding nor retention. Effective teaching requires adding discussion and direct instruction to the mix. Direct instruction usually fails in the absence of inquiry, reflection, and discussion, but it can be powerful in conjunction with them.

In the final wrap-up, the teacher should attend to these crucial endgame concerns.

- *Make key concepts explicit*. It is not sufficient to have used the ideas in context. They need to be stated clearly and unambiguously.
- *Clarify what is important* and worth remembering, and thus worth writing down. At the end of a lesson, the central idea is not necessarily obvious to all students.
- *Help students make connections* with other representations and previous knowledge. This may require further exploration and/or discussion. In any case, it does not often happen spontaneously.
- In addition to a recap of the lesson's key concepts, institutionalization also includes the sharing of *essential conventions of the field*, such as terminology and notation. Those, by their very nature, are not candidates for discovery.

When the teacher summarizes, students should write things down as a reference they can return to. They should write down what the teacher asks them to write, in a place where they will be able to find that summary later. Such teacher-controlled notes are vastly more useful than what students might put down on the fly when the teacher is talking.

## All of the Above!

"All of the above", but not "anything goes!" Our foundational idea is that math teachers should reject shallow either-or stances, and instead *learn to combine multiple approaches*, even if they superficially seem to contradict each other. There is no one way: we need an eclectic mix of techniques that prioritize student understanding. We should avoid the distractions of irrelevant requirements and constant note-taking. The main path to student understanding is intellectual engagement. In our next chapter, we discuss the main vehicle for that engagement: problem solving.

## Discussion Questions

1. How is "embracing contraries" different from "finding a compromise"? Is it always desirable? Is it always possible?[7]
2. Can students engage intellectually while taking notes? If not, how can the teacher help them focus on one or the other at different times? Discuss this in the context of an upcoming lesson.
3. How might you conduct the final "institutionalization" wrap-up in order to dissuade students from simply waiting for you to convey the points of the lesson at the end?

## Notes

1. G. Freeman and L. Lucius. Student engagement and teacher guidance in meaningful mathematics: enduring principles. *The Mathematics Teacher*, 102:164–167, 2008.
2. P. Elbow. Embracing contraries in the teaching process. *College English*, 45:327–339, 1983.
3. mathed.page/teaching/art-of-teaching.pdf
4. Research mathematicians find that simply knowing two definitions are equivalent often brings new knowledge about both. A famous example from astronomy is "The morning star is the evening star." Knowing these are both a manifestation of the planet Venus is incredibly enlightening as to the nature of the heavens. Ditto the particle/wave nature of light, or the apocryphal story about Newton realizing the apple and the moon are both falling objects satisfying the same law and equation.

5. nixthetricks.com — this was compiled by Tina Cardone, with input from many math teachers.
6. A. Arcavi. Symbol sense: Informal sense-making in formal mathematics. *For the Learning of Mathematics*, 14:24–35, 1994.
7. To structure this discussion, see the above-mentioned worksheet: mathed.page/teaching/art-of-teaching.pdf

# 2
# Problem Solving

Problem solving has less to do with teaching specific topics and more to do with how to teach almost anything in the curriculum. If students are to have any idea that math is more than applying procedures, they have to try problems that do not boil down to applying a procedure. That, by definition, is problem solving.

Certainly, there is a place for applying known procedures: it can confirm understanding of those procedures; it can help automate useful techniques; it can provide useful review of formerly familiar material. But if that is students' main involvement with math, they will come to believe that this field is mostly about applying procedures. This is not true about mathematics, and it is not as useful to students' intellectual growth.

Solving problems that you have not (yet) learned how to solve helps build powerful habits of mind, which help students throughout their educational trajectories, and in fact throughout their lives. We briefly discussed some of those habits of mind – the meta-curriculum – in this book's introduction.

A student has *ownership* when they have assimilated the information in a way that makes sense to them. Often this involves some sort of mental housekeeping, as they reimagine the concepts in terms of those they already understand. In the end, the student feels that they could have discovered these ideas themself, or at least could now if they forget them. *Initiative* refers to a student's willingness to try to forge ahead on their own. It is the attitudinal prerequisite for problem solving. It is a result of faith in the student's own facility to reason: the truth is out there, and although the teacher is more knowledgeable, the teacher is not a necessary conduit.

Problem solving promotes both ownership and initiative. Initiative comes closest to bike riding in that it can only be learned by doing. Problem solving always starts with trying something. Ownership is more subtle. A student does not have to have discovered something independently in order to have ownership of the concept. In the long run, though, ownership relies on introspection, questioning whether a thing is really true and what the alternative might be. Discovery-based activities are where this most often takes place. A problem-solving activity is a discovery activity with a specific goal and no obvious path to follow.

There are other reasons to incorporate problem solving throughout the curriculum. It is a good way to keep interest in old, dull, or easy material. One of the challenges posed by classroom heterogeneity is the fact that a given problem may be too difficult for some students, and simultaneously too easy for others. This can be addressed in part by offering problems at various levels. Rich problems work surprisingly well with a range of students: they are easy to start and difficult to completely exhaust. They require exploration, which is the first step of problem solving.

A steady exposure to problem solving does not imply that all topics are approached this way or that it is the primary mode of teaching. We are under no illusion that problem-based lessons suffice to teach the mathematics curriculum in most schools. It needs to be complemented with explicit discussion of concepts, with the generalization of ideas that emerge in the problem-solving process, and with the building of needed skills. Thus, the role of the teacher is essential. Even the best problems do not automatically confer understanding to their solvers. Our claim is more modest: problem solving can and should enhance every aspect of math education.

## Two Sample Problems

This discussion gets too abstract without examples of the kind of material we mean. What kind of problems are we talking about? Here are two examples.

### A Dissection Paradox

We begin with an example, a 19th century puzzle or brain-teaser. Back then, among the educated class especially in Britain, amateur math was a popular pastime. Adults of a certain bent would try their hand at math problems and puzzles published in newspapers. Dissection problems were big hits: how can you cut up a certain shape and rearrange the pieces to form another one, sometimes with limits on the number of cuts? The puzzle we present here is to explain a paradox: what is fishy about a certain dissection.

In Figure 2.1, an 8×8 square is cut into pieces and rearranged to form an isosceles triangle with base 10 and height 13. The area of the square is 64 but the area of the triangle is $\frac{13 \times 10}{2} = 65$. How can this be?

**Figure 2.1** A paradox!

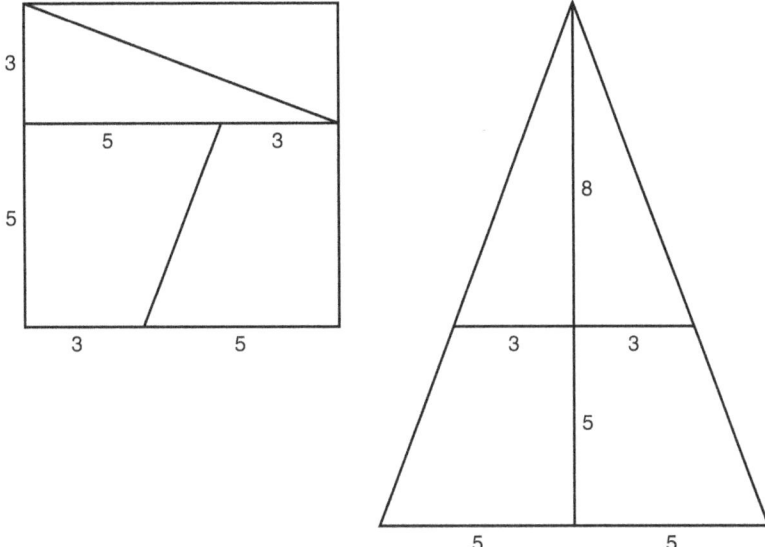

The resulting activity is a rich activity because there are several stages of success. The first is to recognize that the pieces don't fit together. We have sometimes done this activity with graph paper and scissors. Most students don't cut carefully enough to make it obvious that the pieces don't fit, but some do. A student who claims the pieces don't fit together has in some sense solved the puzzle, ruling out other explanations such as that area is not preserved by rearrangement, or the pieces shown are not the same, or that area formulas have somehow been misapplied.

A second stage of success is to say how the pieces fail to fit together. Do they overlap in the middle? Do the angles in the middle fail to add up to 360°? Are the sides not straight? It is the last of these: the slanted sides of the triangle are bent in the middle (the base is perfectly straight – see Figure 2.2). A third stage of success is to prove that the sides are indeed bent. If ADE was straight then the right triangle ABD with legs 8 and 3 would be similar to the big right triangle ACE with legs 13 and 5. This would mean that $\frac{13}{8} = \frac{5}{3}$. False! (Though it's close.[1])

Problem Solving ◆ 25

**Figure 2.2** Similar triangles?

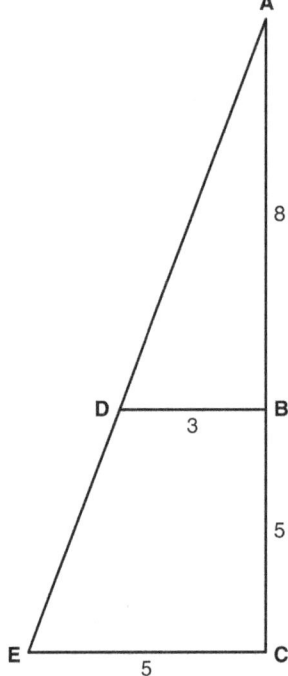

For middle school grades, this is more or less the end of the problem, but at the high school level, we can take it a step further. We can go back to the picture proof that the area of a parallelogram is the base times the altitude.

**Figure 2.3** Is this correct?

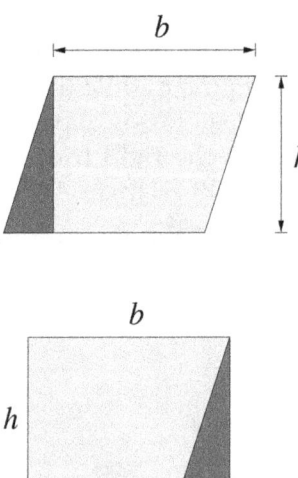

26 ◆ Pedagogical Principles

This involves cutting a triangle off of one side and pasting it onto the other side as shown in Figure 2.3. Now that you have seen the fallacy in the square-to-triangle area puzzle, you should be suspicious about careless dissections! How do you know that rearranging the parallelogram this way actually forms a rectangle? This requires building a logical argument based on the angles involved in this figure.

**Area Versus Perimeter with Polyominoes**

The next problem is a particularly good example of a *low threshold, high ceiling* problem. The phrase "low threshold, high ceiling" comes from the Logo movement in the 1980s. Logo was a programming language designed at MIT to allow beginners to get interesting results early on. However it was a full programming language, and there was no upper limit on the complexity or ambition of student-written programs. (Since then, we've seen variations on this phrase, such as "low floor" and "no ceiling".)

In math education, low threshold, high ceiling problems allow for immediate engagement by the whole class and offer a serious challenge for the students who are fastest to catch on. "Low threshold" means that all students are capable of getting something out of the problem. Usually, this something will be a curricular item. "High ceiling" refers to the existence of multiple further questions that arise, that lie within the range of students with various strengths: to see patterns, or to get a glimpse of some broader phenomenon or larger-scale organization. This can lead to intrigue, articulation of reasons, possible mathematical argumentation, generalization, abstraction, anticipation of future curricular goals, etc. (It also helps students who work quickly through the basic problem to remain engaged.)

If you join unit squares edge-to-edge, you get a shape known in recreational mathematics as a polyomino. (The word is a generalization of domino, a shape made of two squares joined edge-to-edge.) Another way to make polyominoes is to draw closed shapes on grid paper, following the grid lines without crossing your path. Two examples are shown in Figure 2.4.

**Figure 2.4** Two polyominoes.

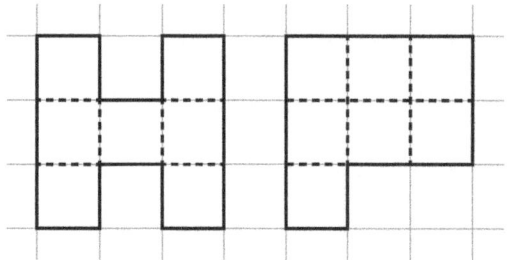

**Problem**: For polyominoes of a given area, what perimeters are possible?

Some experimentation on grid paper reveals that all possible perimeters are even numbers. (Why is that?) After a while, the question can be rephrased: for polyominoes of a given area, what is the maximum possible perimeter? The minimum?

The maximum is not too difficult to find, and most students will find a formula for it. The minimum, on the other hand, is a lot more challenging to sort out. Typically, students find a pattern and are able to predict the minimum perimeter for a large area, for example 1000. However, a formula usually remains out of reach. Figure 2.5 shows a graph of the results.

**Figure 2.5** Longest and shortest perimeter.

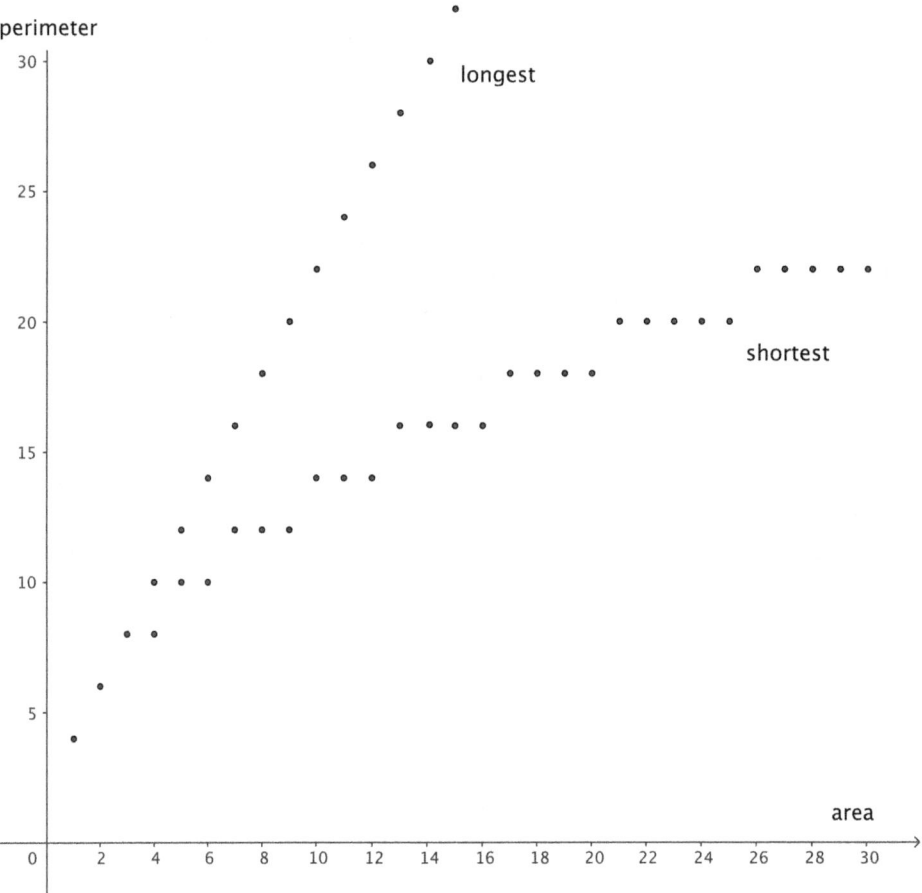

What do students get from working on this problem? The low threshold is particularly evident here. The first half of the problem, up to where the problem is rephrased as finding the maximum and minimum, involves understanding the problem statement and that we are looking for a range. Benefits for all students accrue on two levels.

- Meta-curriculum: concentration, perseverance, generalizing, collaboration, reasoning (for understanding why only even perimeters are possible), and algorithm creation (for predicting minimum perimeters for a large area).
- Curriculum: word problem skills, definition comprehension, comparing a linear function with another function, symbol sense (when comparing possible formulas for maximum perimeter), early exposure to the important topic of optimization, and the quadratic relation of maximum area to perimeter.

In some ways, the problem is "pure math", but when we shared it with an architect, he said that this question was intimately related to things he thought about all the time. In architecture, there is a constant tension between wanting to reduce the perimeter of a building (less expense building walls for the given area, less loss of energy), and wanting to increase it (more windows, and thus more natural light and more view of the outside).

## Placing Problems in the Curriculum

When we say "problem solving", we are talking about problems whose solutions are not obvious. Such problems should not be separate from teaching the core curriculum. Quite the opposite: large and small problems can play multiple roles in the delivery of a curriculum. They can trigger curiosity and motivation. They can help introduce new concepts. They can confront misconceptions and provide opportunities for review and the application of learned concepts and techniques. In what specific ways, then, can non-obvious problems be used?

One way to use problems of this sort is *in the introduction* to a new topic. The dissection paradox can precede formal proofs of the Pythagorean theorem, which often rely on the rearrangement of parts. Or it can follow those proofs if those were initially approached informally. In either case, arriving at a concept twice, once explicitly and once because it arises in a problem, is a big morale booster. The second time the students see it is a spark of excitement that an abstract math concept turned up in real life, or an "Aha!" of recognition that an abstract concept turned out to represent a notion they already understand.

Problems can be used *during* a curriculum unit to give students practice and keep them focused on making mental connections. This also gives the teacher an opportunity to keep the work interesting when there are some students who need more practice and others at risk of getting bored. Problems that work at those two levels are essential if one is to avoid turning math class into a race to reach answers, with some congratulating themselves on their speed, and others giving up and preferring to not even try.

Problems that come *after* a curricular topic also help with practice, but they do more than that. They help students see that they've learned something – they can see that the problem would have been more difficult, or even out of their reach without this new understanding. The prime time for this is just after students have started to master an idea, seeing through new instances quickly and being able to think more broadly without stopping to re-figure out what's going on at every step. The Egyptian Fractions Challenge,[2] for example, could be used in middle school to review essential understandings:

> Write each fraction $\frac{4}{4}, \frac{4}{5}, \frac{4}{6}, \ldots$ as a sum of three or fewer unit fractions (fractions whose numerator is 1). For example, $\frac{4}{5} = \frac{1}{2} + \frac{1}{5} + \frac{1}{10}$.

The Egyptian fractions problem works well once the manipulation of fractions is well understood when even those students who need more practice are not completely consumed by the mechanics.[3]

Another benefit of ending a unit with some increasingly challenging problems is to teach differentially. We owe our strongest students challenges and a glimpse of something deeper. Having a few more challenging problems at the end of the unit that not every student is expected to get to, but that reward those who try them is one way to pull off some differentiated teaching without tracking.

Finally, one might see the need for a problem-based activity to address a topic no one seems to get or to confront a misconception. For example, many students believe the square of a number is always greater than the number, and likewise, that the square root of a number is always less than the number. This comes from being used to integers and possibly from not having much facility or intuition for multiplying fractions. Working on the problem shown in Figure 2.6 should help prepare the class for a needed conversation. Students should be asked to fill out the tables with the areas and sides of squares a, b, c, and d. If the answer is not a whole number, they should enter a decimal rounded to the nearest $\frac{1}{100}$.

**Figure 2.6** Square sides and areas.

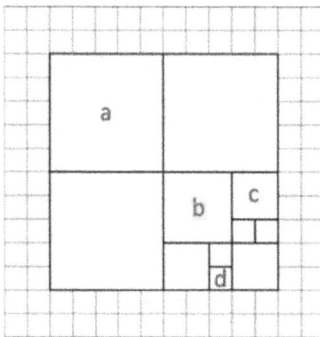

|  | Area | Side |
|---|---|---|
| Total | 100 |  |
| a |  | 5 |
| b |  |  |
| c |  |  |
| d |  |  |

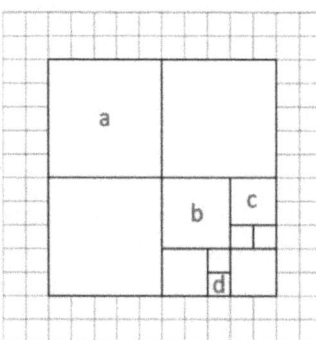

|  | Area | Side |
|---|---|---|
| Total | 1 |  |
| a |  |  |
| b |  | 0.30 |
| c |  |  |
| d |  |  |

## Inquiry

In this section we discuss *inquiry-based* learning. The terms *problem-based* learning, and *discovery-based* learning are also used. Inquiry-based learning can refer to an individual lesson or an entire unit or curriculum.

In most textbooks, an idea or technique is presented, examples are worked out, and students use exercises to practice. This format has a number of advantages, including strict control over pacing, clarity as to what is covered and what is expected, and the use of the textbook as a unifying reference where it is easy to look up needed information. In this model, the teacher serves as a spokesperson for the textbook, passing on correct information to the students; the students' ideas are not deemed relevant except insofar as they do or do not accomplish the task at hand. A disadvantage is that students can often carry out such a program without gaining the understanding that will be needed down the line. Another disadvantage is that the students' thoughts do not enter into consideration except insofar as they are checked for mastery at each step.

At the other end of the spectrum from this traditional approach are problem-based textbooks and curricula. They offer interesting problems, organized in a way that the authors hope will guide the students to important realizations. One example of this is offered online by the Philip Exeter Academy's math department.[4]

Problem-based approaches require students to engage intellectually with significant mathematics. They make it easier for the teacher to see what students actually understand – an essential reality check. The disadvantages of problem-based curricula are the negations of the advantages of traditional curricula: loss of control over pace and the danger of failing to reach agreement on a correct method. Since the concepts are not explicitly stated, it may be possible for a student to complete a discovery activity without grasping the intended punch line, or even without realizing what the activity is about. It is also more difficult to use problem-based textbooks as references, or to integrate parts of their content into other curricula.

Somewhere in between are textbooks that use a *guided inquiry* approach. Such books organize the material in clearly topical units, state important results explicitly, and prepare students for those results through engaging explorations and rich activities. Even if such books have not been adopted by your school, they make excellent references for preparing lessons. Some examples are the Illustrative Math materials,[5] the College Preparatory Math series,[6] and the pioneering books by Chakerian, Stein, and Crabill. Henri came across their *Geometry: A Guided Inquiry* (*GGI*) early in his career. This way-ahead-of-its-time 1970 book, like this one, was the result of a collaboration between a teacher and a mathematician. (Actually, two mathematicians in that case.)[7]

Henri writes:

> This book had an enormous impact on me, both as a teacher and later on as a curriculum developer. In fact, of all the books I have seen in my decades in math education, this is probably the one that taught me the most, by far.[8]

Here are some things we learned from this book:

*Practice need not be boring.* For example, the book had many enjoyable problems in the form "what's wrong with these?" which showed figures which violated one or another theorem. This presents an opportunity to practice important ideas in an entertaining activity. Figure 2.7 features an example.

**Figure 2.7** What's wrong?

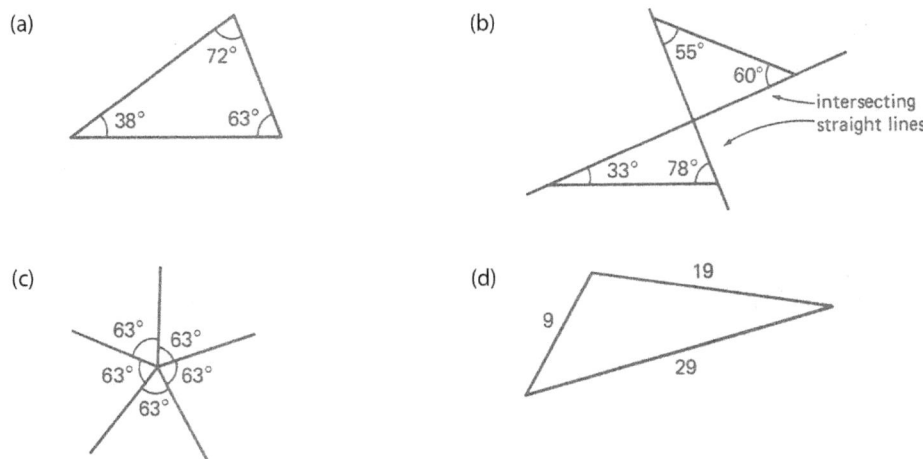

*Problems need not be sequenced in order of increasing difficulty.* This is countercultural but effective. When students don't know if the next problem is going to be easy or difficult, they are more likely to give it a shot. If the problems get harder and harder, many students will reach a point where they decide they can go no further.

*Answer-getting is not the point.* In fact, some answers are given right there in the margin, which allows students to check their understanding as they work. Most other answers are given at the end of each chapter.

GGI showed how guided discovery can provide the right balance between student discovery and direct instruction. Neither is sufficient without the other. On the one hand, students often cannot hear answers to questions they do not have. On the other hand, they cannot re-discover all of math. The key to a healthy combination of discovery and direct instruction is the use of worthwhile problems that are both accessible and challenging, both before and after the key results are presented explicitly.

To summarize: guided inquiry has the advantages of both traditional and problem-based books, and it tends to mitigate their disadvantages. It can help teachers move from a fully traditional approach toward increasing the problem-solving component of their classes. We suspect that on balance it is the most effective approach for most schools.

Another sort of compromise between traditional and problem-based curricula is to insert some problem-based units to complement or replace the corresponding sections in a traditional textbook. This can serve as a transitional stage toward a guided inquiry approach, but it can also be a deliberate choice in order to balance coverage concerns with the importance of problem solving.

Of course, an experienced teacher with a strong understanding of mathematics and pedagogy can successfully use any textbook, or no textbook, to implement a guided inquiry program.

## Rich Activities

Rich activities accomplish multiple objectives. In a tightly packed curriculum, the efficiency created by rich activities is an important way to accomplish meta-curricular goals without giving up content coverage. They have as many of the following characteristics as possible.

- They carry significant *mathematical content*, as opposed to easily forgotten micro-skills.
- They are accessible (*low threshold*): the question is easy to understand, and every student can get started.
- They are challenging (*high ceiling*): there are opportunities to go deep and far – the question is of interest to the teacher as well as to the student.
- There are interesting *partial results*: those who do not go deep and far still learn something of interest and of value.
- They are *engaging*: they trigger student curiosity while offering avenues for exploration.
- They are *student-centered*: while the teacher structures and guides the activity, students do the work.
- There are *many paths* through them – this helps to reduce damaging competitiveness and reinforces ownership.
- They are *group-worthy* – students can work collaboratively and discourse is enhanced.

Note that rich activities provide a way to address the needs of students who work unusually fast: instead of giving them unrelated challenging problems to keep them busy, the high ceiling makes it possible to keep them involved with the same work as their classmates. This helps reinforce the feeling that the class is a learning community, and it gives all students something to strive for.

Here is an example of a rich activity:

Suppose you write down all the products of two consecutive numbers from $1\times2$, $2\times3$, $3\times4$, ..., to $99\times100$.

How many are divisible by 2?
How many are divisible by 3?
How many are divisible by 4?

How many are divisible by 5?
How many are divisible by 6?
How many are divisible by 15?

Each student can make it some part of the way through this, and get mini-aha's all the way along.

Or compare these two approaches to learning about the area of a trapezoid.

### Activity #1

The teacher recalls for the students that the area of a trapezoid is given by the formula $\frac{h(b_1 + b_2)}{2}$, where $h$ is the height and $b_1$ and $b_2$ are the bases. The students are given a practice worksheet containing a dozen examples, each with different numbers for the bases and the height. The more advanced examples may have variables instead of numbers or require students to compute these quantities based on other geometric information given. To make it more vivid, we can imagine a few more details of this activity. The students practice in silence. Some students are happy because they know exactly what to do and are doing it. Other students don't like it, because they find it boring. All of them know they will soon have to calculate some trapezoid areas on a quiz, so they are at least grateful for the practice. Some will make an effort to memorize the formula in preparation for the quiz. Many will have forgotten the formula a week, a month, or a year later. This is because they will not use it again unless they take calculus many years hence. In any case, whether they remember it or not, doing the exercises does not help them understand the formula.

### Activity #2

Students are given a sheet of paper with a few copies of a certain trapezoid on it, with all the measurements indicated, including the bases, the legs, and the height. They are not given a formula (thus it is a discovery-style activity, as are most rich activities). They are asked to find the trapezoid's area. In fact, the teacher has introduced the worksheet before teaching any formula. Students are allowed to use scissors to cut out the trapezoids, and if they want to, cut them into smaller pieces that can be rearranged. Rearranging would allow them to use formulas they already know, such as the one for the area of a rectangle, a parallelogram, or a triangle. (Scissors are not necessary. For example, the activity can be carried out on grid paper, without any cutting. Which is preferable will depend on the specifics of a given class.) Students will almost certainly come up with different strategies, such as those pictured in Figure 2.8.

**Figure 2.8** Trapezoid strategies.

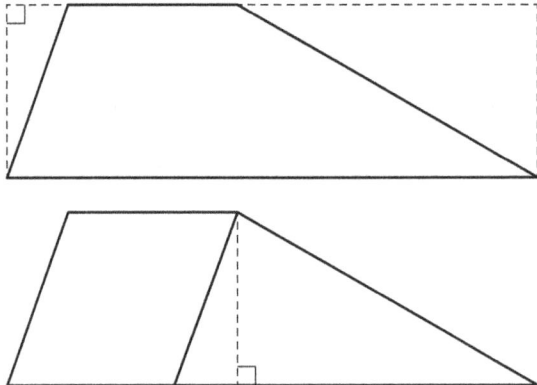

Students who find one quickly can be encouraged to look for more. Some students may not like the activity, because they are not told exactly what to do. The teacher can offer hints to them, or encourage them to get help from neighbors.

Once some strategies have been found, the teacher can lead a discussion where students demonstrate their approaches. Several strategies will reveal that the lengths of the legs do not contribute to the final answer. In fact, all correct strategies will yield the same answer for the area. If some strategies yield a different answer, all the better: a good discussion can be had about what went wrong. A general formula can be the final punch line: applying any of the strategies to a generic trapezoid always yields the same (or an equivalent) formula.

Even if the first approach includes a teacher explanation, we maintain that the second approach is far richer. The derivation has more mathematical content than does the formula. Everyone can get started. Few will immediately appreciate the subtleties: how many ways there are to do this, which ones always work, the relation to formulas and algebra. Activity #2 is student-centered, group-worthy, and we think, engaging.

Not coincidentally, the second activity accomplishes far more of the meta-curriculum as well. To start with, there is longevity. Many students who cannot remember the formula at some point in the future will be able to use one of the strategies that came up in the course of the exploration, either to find a particular trapezoid's area, or to reconstruct the formula. Activity #2 also carries the message that formulas can make sense, that there are many ways to solve a given problem, and that not everything needs to be memorized. A perhaps unexpected bonus is that the different solutions to this essentially geometric problem yield different interpretations of the formula, and some apparently different but actually equivalent formulas.

At the end of the activity, if the teacher has an excellent explanation of the formula that was not found by the students, nothing prevents them from sharing it. Doing it after the hands-on phase of the lesson will mean that students are a lot more likely to listen to and understand the teacher's explanation. Likewise, if practice in using the formula is deemed to be important, it can still be done after Activity #2, perhaps as homework.

Are there any downsides? To be sure, the second approach takes more time. (Part of that time may be spent on algebra when comparing formulas, so the time is not wasted!) In the long run, it is more efficient to teach in a way students will remember, but in the short term, a teacher may want or need to fulfill their responsibility to content coverage in a less time-consuming way, thus passing on the problem to a future teacher. Another downside is that some teachers are simply not used to handling open-ended activities, as they require on-the-spot decisions and skill in delivering the final wrap-up. In that situation, the teacher will need support in transitioning to this style of teaching. For now, we are just making a point about how to best achieve understanding and retention.

## Creating Problems

Unfortunately, few textbooks give problem solving the prominence we advocate, so it is sometimes incumbent on the teacher to find or create suitable problems. Some sources are good if you just want to find some cool problems but are hit-and-miss for coming up with problems on a given topic. Standardized test problems can work well if divorced from multiple choice and time pressure. Sharing among colleagues, whether at one's school or across the Internet, is probably the best way to build a portfolio of good problems. Mathematics competitions are another source, though those problems may be too difficult, and not close enough to curricular goals.

For teachers who have the time, interest, and energy for problem creation and adaptation, we share some strategies.

### Reversing

One strategy is to reverse a standard exercise. For example, the standard exercise, "What is $7 + 3$", may be reversed in several ways.

- What $+ 7 = 10$?
- Find pairs of numbers that add up to 10.
- Find sets of numbers that add up to 10.

Similarly, reversing the standard exercise, "What is the greatest common factor of 12 and 18?":

- Find a number so that the greatest common factor of 12 and your number is 6.
- Find many numbers whose greatest common factor with 12 is 6; what do they have in common?
- Find pairs of numbers whose greatest common factor is 6; what do they all have in common?

Although a steady diet of extremely open-ended problems is too much, an occasional such problem sparks interest. For example, you can reverse a problem such as, "Solve $2x + 3 = 11$ for $x$", by asking students to make up a problem of the same form where the solution is $x = 4$. In Chapter 7, we discuss reversing "graph this function" by asking students to find the function that yields a given graph. This is more interesting and yields more insight.

Reversal provides a powerful mechanism for the construction of problems: start with what you are trying to teach or apply, and reverse the question. Voilà! You have created a problem.

**Other Problem-Adaptation Strategies**

On his website, Geoff Krall proposes a number of additional strategies to improve textbook exercises.[9] Here are three of his suggestions.

- Given a whole page of boring drills in the book, ask students to pick five problems to do, along with an explanation of why they chose them. And five problems that they really do not want to do, along with an explanation of why. The assignment will be popular, for obvious reasons, but it will also get the students to look at the problems in a meta sort of way – a valuable skill, and possibly the germ of some good discussions.
- "Spill coffee" on some of the givens, making them illegible. Students choose their own givens and do their own version of the problem. This will probably be tailored to their level of expertise.
- In multistep problems, remove the intermediary steps. (This often makes the purpose of the problem a lot clearer.) But don't throw them away! Use them as hint cards, or as debrief prompts.

Inspired by classic screenplays, Dan Meyer suggests creating a "three-act" version of a standard problem.[10] Act 1 consists of setting the stage "clearly, visually, viscerally, using as few words as possible", in order to engage every student. Critical information that would be needed to solve it is omitted. In Act 2, the teacher asks what tools and techniques might be helpful, to solve the problem, and what information is needed. Once that has been adequately discussed, they release that information. Solving the problem will almost certainly go better than it would have if it had not been preceded by Act 1. Finally, Act 3 is a debrief of the problem, and setting the stage for a sequel – an extension problem. We might add that this may be a good time for a discussion of any generalizations that can be gleaned from it.

## Before and After

Another strategy is to take a standard exercise but to present it *before* teaching the underlying concept. Students cannot always hear the answer to a question they don't have. A consequence is that even after a great explanation from the teacher, they may not absorb what they have been told. If instead of starting with an explanation, you start with a question to get students to think about the topic, they are much more likely to understand your explanation.

For example, in preparation for the above trapezoid example, you can ask students to find the area of graph paper triangles, as in Figure 2.9, before mentioning that there is a formula for that.

**Figure 2.9** Triangle area.

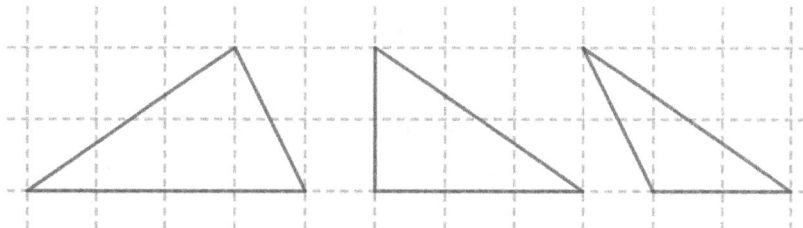

There are many benefits to this. It may reveal strategies that make more sense to the students than the ones you come up with. It gives students the confidence they can recreate the formula if they don't remember it. It makes "base" and "height" concrete and visible rather than numbers on a figure. A discussion of student solutions can lead to a discussion of how to generalize what was learned, leading up to a formula. And of course, starting this way does not prevent you from providing your own explanation later.

Here is another example that is best explored *before* the curricular unit on comparing linear and exponential functions than it is in the middle.

> Bea did so well in algebra that she got a job as an algebra tutor. Her starting salary, as she had no experience, was $10 per week.

As Bea got more experience, her salary increased. She got a raise of $1 per week.

Abe also got a job as an algebra tutor. He heard that Bea was getting a weekly raise of $1. Since $1 is 10% of $10, Abe asked for a weekly raise of 10%.

What will happen in the first ten weeks?[11]

Middle school students can explore this, one step at a time, without having been formally introduced to linear versus exponential growth. (Admittedly, the calculations would get tedious without calculators.) By working this out methodically, they will be much better prepared to discuss these two types of growth. High school students will be able to take this further, perhaps to a general formula for each type of growth.

Another more challenging example is, before teaching how to turn a fraction into a repeating decimal, ask the students to compute some. They can see that $\frac{1}{3}$ repeats, as does $\frac{1}{9}$ and $\frac{4}{11}$. Ask whether $\frac{1}{7}$ repeats. If you want to have some fun, tell them you have a fraction that does not repeat, say $\frac{1}{23}$. Can they figure out why it must repeat, without doing the computation which will be slow and possibly inaccurate? (The argument is based on what would have to happen eventually as you apply the long division algorithm to dividing 1 by 23: there will be at most 22 different remainders, and after those are exhausted, a repeat is inevitable.)

After working on a certain type of problem, the teacher can ask students to create their own problem of that type. For example, ask students to make their own simultaneous equations word problems. Especially in the arena of word problems, it is illuminating to see what ways the problems fail to capture what you asked them to capture. A well-known study asked preservice teachers to create a word problem illustrating $6 \div \frac{1}{2} = 12$. The success rate was surprisingly low.[12]

**Problematic Problems**

Problems that are a little off help students learn argumentation. Was there not enough information, or contradictory information? Was the problem better done by methods unrelated to what the problem was supposed to be about? Do the numbers make sense in the real world, and are the numbers in the problem chosen so that the solution is plausible? Here is one approach, suggested by Scott Farrand, a professor emeritus at Cal State University in Sacramento:

> In some algebra classes, most of the students simply do not engage with word problems, perhaps because of bad experiences in the past, because of the reading load, the density of information, the

weird language used in math word problems, or because they had been taught ways to solve problems without reading them.

Here is a suggestion: When giving the students a word problem, first give them a version of that problem that has been screwed up, usually a number changed, so as to make it not make sense. For example, if the problem is altered to say that someone has nickels, dimes, and quarters in their pocket, twice as many nickels as dimes, and one more quarter than nickels, and the total value of those coins is $3.27, then the students can call BS on that problem and they understand a little about the context of the problem. They enjoy finding fault with what the teacher has presented. And they can tell you one thing that would have to be different for the problem to possibly make sense. Once they do that, you can give them the real problem, by revealing the total value of the change.

This sort of thing is remarkably easy to do with many of the standard word problems. It is an easy routine for word problems that makes them more inviting to students and that helps them find a way in.

**Non-Random Drill**

It is important to distinguish problems (which require thinking), from drills, which are intended to help students master and automate various skills. Drills of important skills can be useful, but if handled poorly, they can be counterproductive. If drills are the only or the main activity in a math program, they give the wrong impression about what math is, whether students enjoy them or not. If they are divorced from meaning, and seen as pointless, they undermine students' disposition toward math.

One way to make drills more interesting is to use non-random exercises. By this, we mean sets of exercises which require the practice of needed skills while illuminating interesting patterns. Such sets often allow students to chart their own path through the possibilities. One example is the Egyptian Fractions Challenge we mentioned above. Here are two more examples.

- Instead of a page of arbitrary multiplications, ask fourth graders to fill in the blanks in the expression (Figure 2.10), using the numbers from 1 to 9 at most one time each. The goal is to get the largest product possible.[13]

**Figure 2.10** Maximize!

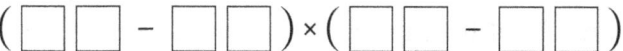

- Geometry students could be asked to find the distance of lattice points (points with whole-number coordinates) from the origin, perhaps limited to points whose coordinates are between 0 and 10 inclusive, expressed in simple radical form. This is a way to practice the Pythagorean theorem and the manipulation of radicals, but the fact that it is non-random makes it possible to notice patterns and find shortcuts. For example, the distances for (1, 1), (2, 2), (3, 3), etc. are all multiples of $\sqrt{2}$.[14]

In short, non-random drill injects an element of problem solving into the needed practice, which makes the work both more interesting and more effective at getting concepts across. In these examples, drill is in the context of an interesting overall quest, and thus much more motivating. Also, unlike random drills, it lends itself to reflection, discussion, and generalizing.

## Using Puzzles

Recreational mathematics offers many problems and puzzles that lend themselves to classroom use. Martin Gardner is the towering figure in this domain. In the preface to his book, *Mathematical Carnival*,[15] he compares his own work in writing columns about math for the general public to the work of a math teacher:

> Surely the best way to wake up a student is to present him with an intriguing mathematical game, puzzle, magic trick, joke, paradox, model, limerick, or any of a score of other things that dull teachers tend to avoid because they seem frivolous.
>
> No one is suggesting that a teacher should do nothing but throw entertainments at students. And a book for laymen that offers nothing but puzzles is equally ineffective in teaching significant math. Obviously there must be an interplay of seriousness and frivolity. The frivolity keeps the reader alert. The seriousness makes the play worthwhile.

You may think that a discussion of puzzles strays too far from the curricular concerns of math educators. It does not: a puzzle constructor's ethos is helpful to teachers and curriculum developers. We hope the following discussion will convince you that we are right. Almost everything we will say about puzzles applies more generally to curricular math problems.

A good puzzle is meant to be solved, so everything the solver needs should be there. But if the puzzle offers some resistance, the solver's satisfaction is that much greater. Of course, if a puzzle is overly difficult, frustration and resentment may ensue. Fortunately, a teacher can see whether a student is reaching that point, and can offer hints.

A puzzle is a relationship between the puzzle constructor and the puzzle solver. There is an unwritten contract between the two. Here are some of the contract's clauses:

- The puzzle must be solvable and fair.
- The puzzle must be challenging.
- The solution must be satisfying.

Of course, this applies to any problems we present to our students!

These requirements depend on context. The same puzzle may be too easy to be satisfying for one solver, while another solver might deem it unsolvable, and yet another may consider it "just right". Still, these guidelines may be helpful to puzzle constructors (and curriculum creators) as they provide a way to think about this. For a puzzle to be solvable, it must be possible to imagine some path to the solution. Fairness is harder to determine, as it depends on matching the puzzle difficulty to the solver's probable skill and experience. (For example, something may be an interesting puzzle in 6th grade, but straightforward to an algebra student.) What complicates matters is that insufficiently challenging puzzles are not as satisfying to solve. The purpose of a puzzle is for the solver to "win", but not to win easily.

All of this applies to the use of problem solving in math education, except that in math education, the stakes are higher! Part of the attraction of puzzles is that they are there for entertainment, and not much hangs on whether one succeeds in cracking the puzzle, or not. But this is precisely an argument for the use of puzzles in the classroom: they help lower the emotional stakes while offering a path to engagement for a wider range of students. (Thus they are not good candidates for use in assessments!) Good math puzzles share many characteristics with the rich activities we discussed above, and in fact, in some cases, they can provide the substance of such activities.

Here are some ideas that may help in the creation of good math puzzles.

- The puzzle should be *interesting* to you, the constructor, even if you consider it easy to solve. If you're bored, the solvers will be bored.
- You should mentally inhabit the mind of the solver, and imagine how they might get to the solution. If there are *multiple paths* to the answer, all the better. If there are partial solutions along the way, those help to keep the solver engaged. Alas, not all puzzle solvers appreciate partial solutions. Some completists would rather not ever have tried a puzzle they cannot fully solve…

- You should also try to imagine what a frustrated solver would feel if they broke down and looked up the answer. Would they think "Darn, I should've gotten that", or "How did they expect me to figure this out?" The first reaction is the one you want.

Note that students' willingness to tackle challenging puzzles and persevere will largely be determined by their trust that puzzles they have seen before in this class could actually be solved by students. Thus, it is important to proceed with caution, raising the challenge level step by step, and offering hints as needed.

### An Example from Another Domain

Admittedly, those ideas are abstract and general. We will try to make them more concrete with an example from cryptic crossword construction. (Both of us enjoy solving those word puzzles, and one of us creates them.) The crossword structure already allows for multiple paths, as solvers can decide the order in which they solve the clues. In a cryptic crossword, each clue consists of two parts: the definition of the answer, and some other way to get to it, involving some sort of wordplay. In other words, each clue is its own mini-puzzle. This means that there are different entry points for solvers with different skills and backgrounds. Moreover, each individual clue contains three paths to its solution. For example, consider this clue:

Tech pioneer : "I know A – Z, but in a different order" (7)

(The 7 indicates that you're looking for a seven-letter word.)
Let's say that you already have these two letters in the diagram:

W _ _ _ _ _ K

The unusual letters at the start and end of the entry may suggest the answer. Or, you may get it from the definition of the answer ("Tech pioneer".) Or, you may get it from the wordplay part of the clue "I know A–Z, but in a different order", which to a solver of cryptic crosswords suggests anagramming (rearranging) the letters IKNOWAZ. One of those three paths to the answer, or more likely a combination of two of them or all three, will lead you to the solution: WOZNIAK.

When considering this clue, a constructor of cryptic crosswords would have some choices. For example, the solution could be easily researchable: replace "tech pioneer" with "Apple founder". But that would not be satisfying to the solver: whether they already know it or look it up, the answer is obvious, and they would not need either of the other two paths to the answer. Or, instead of "but in a different order", the clue could just say "anagrammed", but that too is just too blatant, and moreover it would take away from the humor of the clue, which is part of what makes the solution satisfying. This clue hits the sweet spot and satisfies all the guidelines we suggested above.[16]

But we should get back to math education. Constructing puzzles for the classroom brings with it additional complications and challenges.

**Features of Effective Classroom Puzzles**

A characteristic of all classrooms is that they are constituted of students whose backgrounds and talents vary widely. Offering puzzle sets can help, as it allows students to find their own way through the set, by selecting puzzles at the appropriate level of difficulty, and/or by pursuing partial discoveries. This addresses classroom heterogeneity while having all students work on closely related problems.

As a particular implementation of rich activities, puzzle sets allow the students to find their own path through them. They avoid a common pitfall of curriculum development, which is the hubristic belief that one is capable of writing a single sequence of puzzles that will work just as well for all students. This is a common failing of both traditional and contemporary curricula.

For example, the consistently brilliant desmos.com environment offers teachers and curriculum developers the ability to craft one-path-fits-all sequential lessons in the Activity Builder. The best Activity Builder lessons incorporate many excellent puzzles. (See for example Marbleslides, which we discuss in Chapter 7.) This is vastly better than most supposedly "intelligent" educational software, which tries to eliminate the need for teachers and is often based on "memorize-the-algorithm-and-practice". Still, one can hope that a future version of the Activity Builder will allow the creation of choose-your-own-path activities.

In addition to the availability of multiple paths, effective classroom puzzles also share other properties.

- They are part of a set of related puzzles, rather than one-of-a-kind puzzles that rely exclusively on "aha" insights. Therefore, solving some of the puzzles helps the student develop skills and intuitions that can then be applied to other puzzles in the set, and more importantly, contributes to their mathematical maturity. This also means that they provide an excellent environment for teachers to provide hints, and scaffold student learning. For example: "solving this easier puzzle will help you make progress on the one you that is currently frustrating you".
- They are interesting to both kids and adults. We have used some puzzle sets in the classroom with students at various levels, and in professional development sessions for teachers, and found that they are just as engaging for all. This is in part due to their "low

threshold, high ceiling" quality: all include simpler and more difficult puzzles. Moreover, they suggest additional questions, such as the creation of similar puzzles, or the generalization of results, or the need for a proof.
- They involve significant mathematics and carry a substantial curricular load. They are about the math teachers and students already know they should teach and learn. Using non-math puzzles as a "change of pace" is a waste of precious class time, and gives students the wrong impression that "normal" math is no fun.

One cannot expect all these criteria to apply to every classroom-bound puzzle or puzzle set, but hopefully they are helpful guidelines for teachers and curriculum developers.

We will discuss puzzles as learning tools, and offer examples, in Chapters 5 and 6.

## Problem Solving at the Core

Prioritizing problem solving, in combination with the ecumenical approach we presented in Chapter 1, is the basis of our approach. Problems are the core of math instruction and should be integrated throughout. Well-chosen problems help students to engage intellectually, which is the key prerequisite to understanding. They also help students develop curiosity and perseverance, qualities which will serve them well in math class, more generally in all school work, and in life. The purpose of this book is to share ideas on how to make this happen in the classroom.

One thing we need to make clear is that while we give a central role to problem solving, we in no way suggest that the teacher should refrain from supporting their students as they engage in it. Quite the opposite: as you will see when we discuss group work, whole-class discussions, and throughout the book, we encourage you to give appropriate hints as needed. Here are some of the sorts of hints we find helpful:

- Try it with small numbers.
- Try it with a numerical example.
- Make a guess, check whether it works, adjust it.
- Draw a diagram.
- How would you recognize a correct answer?
- Try a simpler problem (e.g., one fewer variable, a simpler function, remove a constraint).
- Maybe what you think is true is not true – can you find a counterexample?
- Are you sure you remember what the problem is? What are you asked to do and what were you given?

Among these, you may have recognized standard problem-solving techniques. Mostly, we do not tend to use these as something we teach our students separately. Rather, they are part of our instructional toolbox: something we reach for in context, when students are stuck. Over time, they will pick up those habits, and will not need us to make those suggestions as often.

## Discussion Questions

1. Work on some of the problems we shared in this chapter. How do you think they would play out with students? And/or: try them with your students, and discuss how it went.
2. Practice the art of problem enhancement by choosing problems from your textbook that could be improved by applying some of the ideas in this chapter. Try them with your students, and discuss how it went.
3. A student (or their parent) complains, "It's not fair for you to ask me to do a problem you haven't taught me how to do". How might you respond?
4. What do you usually do when you see that many of your students are stuck on a problem from the start? What are some alternatives to that approach?
5. Have you used or come across "low threshold, high ceiling" problems? Rich activities? Non-random drills? Share them with your colleagues and discuss ways to integrate them into your courses.
6. When attempting puzzles and unfamiliar problems, the students' success rate may drop considerably. How might you handle the sense of failure that some of these students might have?

## Notes

1. An extension is to point out that 3, 5, 8, and 13 are consecutive Fibonacci numbers. Is that a coincidence? (No!) We have used this with 11th and 12th graders, asking for versions of this paradox using other consecutive Fibonacci numbers. It turns out that as the numbers get larger, the ratios approach the Golden Ratio, and moreover, there is an interesting formula involving the product of the outer numbers and the product of the inner pair.
2. mathed.page/early-math/egyptian.pdf
3. This problem is based on the Erdös-Straus conjecture, which states that this decomposition is always possible. This result has not as of yet been proven, but it has been verified well beyond where any

student will reach. We call it "the Egyptian Fractions Challenge" because unit fractions were the only fractions used by the ancient Egyptians.

4. exeter.edu/mathproblems
5. illustrativemathematics.org
6. cpm.org
7. Chakerian, Crabill, and Stein also wrote books for Algebra 1, Algebra 2, and Trigonometry. Those were not as successful as the geometry book, but they contained many excellent ideas, including the 10-centimeter circle we discuss in Chapter 5, and a geometric approach to complex numbers. At the height of the math wars in the 1990s, Henri was involved in a discussion with Chakerian. It ended well. Read about it at mathed.page/geoboard/geoboard.html
8. Henri wrote a review of *GGI* and what he learned from it. blog.mathed.page/2019/03/28/geometry-a-guided-inquiry/
9. emergentmath.com/2015/04/16/nctm_adaptation
10. danmeyer.substack.com/p/the-three-acts-of-a-mathematical
11. This problem is from *Algebra: Themes, Tools, Concepts* (mathed.page/attc), by Anita Wah and Henri Picciotto.
12. Liping Ma. *Knowing and Teaching Elementary Mathematics: Teachers' Understanding of Fundamental Mathematics in China and the United States*, volume 13 of *Studies in Mathematical Thinking and Learning Series*. Lawrence Erlbaum Associates, second edition published by Taylor and Francis (2010), New York, 1999.
13. We found this exercise on openmiddle.com, a site with many problems of this type, contributed by teachers and organized by grade level.
14. This problem is from Henri's *Geometry Labs* (mathed.page/geometry-labs).
15. M. Gardner. *Mathematical Carnival*. MAA, Providence, second edition, 1989.
16. For more on cryptic crosswords, see leftfieldcryptics.com/#how-to

# 3
# Different Modes

Classroom discussion is an essential complement to every other teaching mode because it involves explicit leadership by the teacher, a necessary part of instruction. This is why we devote Chapter 8 to the effective running of class discussions. Still, it is crucial to set the stage for such discussions. Every math teacher has had the experience of explaining something with the utmost clarity, only to find out that their brilliant presentation somehow did not penetrate the students' minds. This can often be remedied by setting the stage with an appropriate activity in another format.

This chapter is devoted to such additional formats. In most, the teacher's role is a mix of listening, prodding, asking questions, and guiding. We share a collection of techniques – some that are part of our own teaching, and others we learned about from various sources. In all cases, we hope that they can be learned without expert guidance, and that they will bring rewards whether they are applied in isolation or as a part of a bigger methodological overhaul.

In many secondary school math classes in the US, there is a fairly rigid routine. First, "going over the homework", which consists of the teacher answering questions, or students writing answers on the board. Then, the teacher presents a new procedure. Then students practice the procedure in silence, perhaps starting the next homework in class. Many administrators, teachers, parents, and students believe this is the way math teaching is done, and do not trust that students can do the sort of intellectual work we promote in this book.

This approach may be appropriate on occasion, but as a daily routine, it is problematic. It is boring. Also, it is intimately related to the misconception that learning math consists of memorizing procedures that are fed to you by the teacher. As it turns out, this is not the essence of learning math. Procedures can be programmed into devices. Students can do better than that. Our goal should be to equip students to tackle a wide range of quantitative and visual problems effectively, and thus we need to use some other teaching modes, modes which support students' intellectual engagement.

We start by comparing formal and informal modes and then describe specific techniques.

## Classroom Basics

There are many, many ways to manage your classroom, and no one choice is always the right one. Sometimes, we should rely on a trusted and familiar routine; but sometimes it is important to go for variety and a change of pace. Sometimes we should use technology or manipulatives, but sometimes students should work on paper, or do "mental math". And so on. Growing as a teacher is all about increasing our repertoire of techniques.

Still, there are some overarching realities that dictate limits outside of which it is not possible to be an effective teacher.

### Classroom Management and Teacher Leadership

No matter what the current lesson plan entails, it is essential that the teacher be in charge. This is often called "classroom management", which makes some sense, since the teacher needs to manage the classroom. But those words may carry an implication that the goal is mostly student compliance. As we see it, the goal of classroom management is to facilitate student learning and teacher leadership.

Our students have a lot going on in their life. Learning math, for many of them, is low on the list of priorities. Moreover, even those students who are eager to learn don't necessarily know the best way to accomplish that. Thus, the teacher needs to be in control. Sometimes it means micromanaging ("get your book out", "write this down", "pull your chair in", etc.) Over time it means letting go and giving students the space to take responsibility – but the only way to get there is to build healthy work habits.

### Formal Versus Informal Atmosphere

There are some societal expectations of classroom decorum. A teacher cannot take a nap. A student cannot burst into song. A visitor cannot make a phone call. And so on. But within that, there is a range of possibilities. Instead of searching for a happy medium, it is best to clearly delineate two distinct modes.

The *formal mode* is reserved for whole-class discussion, for teacher explanations of various sorts, and for quizzes and tests. When in formal mode, students stay in their seats, do not go to the bathroom, and do not speak until they are called on. When the teacher or a fellow student is addressing the class, they listen respectfully. If some students are distracted, the teacher waits until everyone is focused.

However, insisting on that sort of formality at all times can undermine the comfortable atmosphere that is most conducive to productive reflection, small-group discussion, and collaboration.

The *informal mode* is reserved for times when students are working individually or in groups, or doing an activity using technology or manipulatives. They need not be silent, as long as their conversation is about the work at hand. They can go to the bathroom if they must. They can ask for help from teacher or peers. Of course, how relaxed the atmosphere can be at those times depends on the teacher's personality and experience, as well as the school and department culture.

It is much easier to transition from a formal to an informal segment than it is to go in the other direction. Teachers come up with various ways to signal that individual and group work should cease in order to start a whole-class discussion: flashing the overhead lights, ringing a bell, hands clapping, and so on.

In both modes, students are expected to focus on math the whole time. If they stray, the teacher can gently and supportively put them back on track. In particular, if they try to ask irrelevant questions ("is there a quiz this week?", "can we do something fun tomorrow?", "where did you go to college?") the teacher should not be derailed. They should tell the student to meet them after class – and proceed with the lesson.

Of course, each teacher needs to adjust these guidelines to the realities of their personality and their departmental and school expectations. But once they have some clarity on this, they need to make their policies clear to the students.

One way to set up those structures is to ask, early on: "What do you need from your classmates to make it possible for you to learn math in this class?" Typically, students have reasonable ideas: "they should not distract me"; "they should be willing to help others"; and so on. The resulting guidelines can be summarized, shared, and enforced by the teacher. Another approach is to start with teacher-created norms and lead a discussion of the reasons for those rules and expectations.

**Why Norms Matter**

The whole point of such behavioral norms is to make it possible for students to engage intellectually. "There is no royal road to geometry". Intellectual engagement is the only way to learn math, but it is not possible for most students if the teacher abdicates their leadership role.

As teachers, it is tempting to shield ourselves from reality by doing most of the talking. We ask "Are there any questions?", and hearing none, we deceive ourselves into believing that the students are listening to us and understanding what we are saying. One way we extend our self-deception is by emphasizing and prioritizing the memorization of various techniques instead of problem solving and multiple representations. Such memorization can temporarily mask a lack of understanding.

To teach effectively, we need to constantly get a sense of what students actually know and can do. Are they thinking or just parroting? Some of that depends on what materials we use. But some of it has to do with how we teach. Much of this book is intended to help with this "how".

## Openers

A common topic of conversation among math teachers is the daily startup routine. An alternative to going over the homework is to ask students to work on a *warm-up* or *do now* problem, which is posted on the board as students enter the class. This is in part a classroom management device, and it can serve as a way to review previously studied skills and concepts.

### Think First

Scott Farrand proposes a different sort of warm-up, which he calls "Think First". The phrase can be interpreted in more than one way, all of them relevant. He suggests that warm-ups should be challenging (or at least interesting) problems, tightly focused on the day's lesson. The intention is to reveal a key idea, to generate curiosity, or in some cases to dispose of a necessary digression up front. Students are not being assessed, and they can discuss the problems with their neighbors.

Farrand shares this example:

Here are some points in the plane:

| (4,1)    | (17,27) | (1,−5) | (8,9)    | (13,19) |
| (−2,−11) | (20,33) | (7,7)  | (−5,−17) | (10,13) |

Choose any two of these points.
    Check with your neighbor to be sure that you didn't both choose the same pair of points.
    Now find the rate of change between your first and second points.

Students are stunned to learn that everyone in the class gets the same slope. This sets the stage for proving that the slope between any two points on a given line is always the same, no matter what points you pick.

This is an example of using a warm-up to generate curiosity, and it is sure to make for a vastly superior level of engagement in the day's lesson.

Farrand has found several benefits to this kind of opener:

- It has students thinking right at the start of class, which transforms the atmosphere to one of intellectual engagement.
- It is a great lesson-planning tool, as it helps the teacher think of what the key idea is for this particular lesson.
- Sharing the warm-ups with colleagues who teach other sections of the same course is "subversive", as it helps move the culture in their classrooms with little effort on their part. As they circulate and see evidence of students' engagement and understanding, they cannot unsee it.

In fact, this is a great professional development tool, as it puts change within reach: just coming up with five minutes' worth of exploration will get a teacher started on the way to a more problem-rich classroom. That is more likely to happen than somehow making huge changes all at once.

Farrand dismisses the concern that if the problems are too hard, students will not do anything. He points out that if the warm-up is too hard, the teacher can provide hints, or offer an easier version of the problem, either to individuals or small groups, or to the whole class. On the other hand, if it's too easy, they have wasted precious time and gained nothing.

One practical point he makes is that if he writes the problem on the board, some students will spend the whole time just copying it down. Better hand it out on paper, and circulate to make sure everyone gets started right away. Another practical point is that the impact is much reduced if you don't do the warm-ups at the start of class, as by the time you get to it students may have settled into a passive attitude.

**Mental Arithmetic**

Another choice of opener is the Head Problem. This is a brief mental arithmetic or algebra problem or series of such problems. It may concern something the students recently learned. It may be review from longer ago, a mini-spiral in the curriculum. It may pave the way for the lesson about to come. Or perhaps it is an intriguing standalone quickie. For example, students can be asked to mentally calculate $6 \times 54$, $7 \times 39$ as a way to apply the distributive rule. Or to solve $3x - 2 = 43$, as a warmup to a lesson on solving more challenging equations.

The highly polarized debate on calculator use versus paper-pencil computation can be circumvented by incorporating mental math into the curriculum. The way to do this is not via curricular units on mental math, but rather by interspersing mental math into all units. It does not have to be at the beginning of the lesson, but this is a logical place for mental math, for the same reason that Think First problems make good openers: the problem involves active engagement from all students, setting the tone for the rest of the period.

Another advantage is cultivating concentration, the ability to hold items in working memory while at the same time remembering the overall problem you are trying to solve. Finally, head problems make you learn different things. When adding decimals, you typically add the whole part and the decimal part separately, which makes "align the decimal points" obvious and meaningful. Even a one-digit number times a two-digit number is too much for most students to do if they have to reproduce the standard computation in their brains. Learning to break it down in the easiest way is a new skill. When multiplying in your head, you are constantly aware of adding or subtracting a small correction to a big, initially computed, round term.

**Number Talks**

Problems involving numerical computations are given one at a time. Students are asked to think quietly, then give a signal when they have an answer and a strategy. A few students are then selected to share their strategies. As a warm-up, number talks have similar benefits to head problems. A difference is that the emphasis is on strategies rather than answers. A different set of students can contribute, namely those who can articulate strategies rather than those who are confident in their computational accuracy. Problems can be chosen so that multiple pathways are inevitable. Reinforcing belief in multiple pathways at the beginning of a lesson sets an important precedent for the rest of the time. Here is an example of how it might play out in 3rd grade:

**Teacher:** What is $3 \times 19$?
(Students think, and signal when they are ready. Teacher calls on students.)
**Student 1:** I know $3 \times 20$, so it would be one less, so 59.
**Teacher:** I see some of you are disagreeing quietly. Lisa?
**Student 2:** It would be 3 less! So 57!
(Many students signal agreement)
**Teacher:** Did someone do it a different way?
**Student 3:** I know $3 \times 15$ is 45, but then I didn't know what to do.
**Student 4:** I know $3 \times 10$ is 30, and $3 \times 9$ is 27, so 57.
**Teacher:** Can we help Jamal finish his way? $3 \times 15$ is 45, who can help?

## General Routines

These can be used pretty much any time.[1]

### Think, Pair, Share

This is also known as "turn and talk". It has become very popular not only in K-12 classes, but also in higher education circles, where it can be made to work even in very large classes of the size one finds at some universities. Students are given a problem. They are supposed to first think about it individually, then turn to a neighbor and exchange thoughts. The final phase is for a few of them to share their thoughts with the whole class. This routine allows for some peer interaction even in a setting where small group work is not feasible. Also, because the students voice their answers aloud during the pairing phase, the instructor(s) can to some extent eavesdrop and select students to share complementary rather than identical thoughts.

As a rule, one learns more by talking through a problem than by listening to someone else. This basic principle underlies much of what we have to say in the next several chapters!

### Poll the Class

This is exactly what it says. Everyone is asked to respond to a given problem, possibly one that is more involved than those given during warm-ups. Responses can be gathered the way they are for head problems and number talks, or in more elaborate ways. For example, students could be given whiteboards or colored index cards to display answers. In the collegiate setting, instructors often use clickers (or, increasingly, phone apps) for this purpose. Obviously, the idea is not that "majority rules". Rather, the purpose is student involvement, but more than some other routines, this one can be used to get an accurate picture of whether the class is where you think they are (the idea is that everyone is supposed to respond).

McCallum *et al.* suggest that this be used in contexts emphasizing estimation. When you ask for an estimate or a bound (a guess you are confident is too low or too high), it is reasonable to demand an answer from everyone. Making estimation polls a regular feature reinforces number sense and works well in conjunction with mathematical modeling exercises.

During the COVID-19 pandemic, James Propp came across a version of this which is suitable for teleconferencing software, and he still uses it now that he's back in the in-person classroom. He sets up a video meeting at the start of class, but uses only the chat function. Every once in a while, he asks a question of the class, and gives the students time to enter their answers in the chat, using their smartphones. After a while, he asks them to share their answers all at once. He calls this a *chat storm*.[2]

**Contemplate Then Calculate**

Grace Kelemanik and Amy Lucenta, of the Boston Teacher Residency, developed this routine, which is a highly structured format which helps both teacher and student to shift their focus from quick calculation to reflection about structure.[3] It works for many topics. For example, in algebra, the routine requires students to take some time to contemplate an equation such as this one, before attempting to solve it:

$$6(x+2)-4(x+2)+1=21$$

Some reflection (and possibly discussion) helps students see the structure and makes the equation much easier to solve than it would be if they quickly jumped into memorized strategies. Implementing the routine frequently is a great way to help students develop good habits and useful skills, while at the same time communicating a meta-curricular message: thinking usually beats not thinking.

**Kinesthetic Activities**

One way to break up the routine in math class is to have the students get up and experience some of the concepts in their bodies. This is of course helpful to kids who like to move, but it helps everyone make connections and remember things. It can also give a useful reference point when teaching a new concept. The change in venue (from the page to the three-dimensional world we live in) helps to overcome some students' passivity, and triggers potentially valuable deliberation. Of course, such activities are intended to supplement, not replace other work. Moreover, you should not expect miracles from them: while they add a lot to your program, they are most effective if combined with discussion, reflection, and work on paper.

Here is an example, useful for a geometry class, about the idea of distance. These activities are best done in a gym or playground, prior to discussing the corresponding ideas more formally.

1. Choose a student to be Point A. Ask the others to stand so that they are all at the same distance from Point A. Hopefully, they'll make a circle. Introduce the term *locus*: the locus of points at a given distance from A is a circle with A as its center. ("Locus"

is the mathematical term for "location". You can also just say "location" and wait for a future class to introduce "locus".)
2. Choose two students to be points A and B. Ask the others to stand so that they are
   - closer to A than to B,
   - closer to B than to A, and
   - equidistant from A and B.

Explain the word "equidistant": at the same distance. That word is a lot easier to explain now, in contrast with "closer to A" and "closer to B". Initially, students may think that the midpoint of the segment AB is the only point equidistant from A and B, but soon they realize that there are many points that satisfy that condition and that they all lie on a straight line. Discuss the properties of that line (it is perpendicular to the segment AB, and passes through its midpoint. It is the *perpendicular bisector* of AB.)

This activity does not take very much time, and it is vastly more effective than defining those terms at the board and hoping the students were listening. It can be followed up with a formal proof of the properties of the perpendicular bisector.[4]

## Games

Well-chosen puzzles can enhance a math class, because puzzle-solving mirrors the process of doing math. Games, on the other hand, should be handled with care, as students may focus more on winning versus losing than on the underlying math. Also, if you are unaware of how long it will take to explain and complete a game, you might end up with a time sink that's double or triple what you intended. For games to be pedagogically useful, it is important to keep the focus on the analysis of the game.

Consider these two dice games:

> To play, roll a pair of dice 20 times. After each roll, add the numbers on the uppermost faces.
> **Game One**: If the sum is 3, 5, 7, 9, or 11, Player A wins. If the sum is 2, 4, 6, 8, 10, or 12, Player B wins.
> **Game Two**: If the sum is 5, 6, 7, 8, or 9, Player A wins. If the sum is 2, 3, 4, 10, 11, or 12, Player B wins.
> Keep track of how many rolls each player wins. Whoever wins the most rolls, wins that game.

When students play these games, it quickly becomes apparent that Game Two is unfair: Player A will win much more often than Player B. This is surprising to many students, and it triggers a conversation about what "fair" should mean in this context. A detailed analysis of the possible outcomes of rolling two dice makes the difference between the two games clear, and is a good introduction to important ideas in probability.

## Diversifying One's Portfolio

We have argued that there are many alternatives to the traditional explanation/silent practice routine. Effective instructional modes have two common features:

- Active student intellectual involvement in their own learning.
- Sophisticated and flexible teacher leadership.

It is a mistake to put all one's pedagogical eggs in one tactical basket. As long as these two ingredients are present, multiple teaching modes can be used in one class at different times, depending on the nature of the lesson and the available curricular materials. Students do appreciate some predictable routines, but too much of that can take the life out of a class. Switching modes can contribute to welcome changes of pace, and bring some variety to the program. In particular, in a school schedule that involves long periods, changing modes is an excellent way to make good use of the time.

We flesh out our presentation of classroom modes in the next five chapters.

## Discussion Questions

1. How should behavioral expectations differ between formal and informal times in the classroom? Is there even a need to make such a distinction? How can this difference be best communicated to students?
2. Share some classroom routines that have worked for you. (They could be among the ones we shared in this chapter, but they don't have to be!)
3. Which of your own behaviors require adjusting in order to convey behavioral norms you wish to inculcate in your students?
4. Have you had success with using games in the classroom? If so, how do you make sure they work to communicate math content?
5. "I can't do math in my head". How might you encourage or motivate a student who can't or won't participate in mental arithmetic?

**Notes**

1. A number of these are so-called General Instructional Routines taken from McCallum *et al.*, in the teachers' guide to their curriculum *Illustrative Mathematics* (illustrativemathematics.org). Some are in turn attributed to the *Mathematical Language Routines* developed by the Stanford University UL/SCALE team (ell.stanford.edu).
2. mathenchant.wordpress.com/2022/10/15/teaching-with-magic-paper/
3. G. Kelemanik and A. Lucenta. *Habits of Mind*, 2015.
4. For more kinesthetic activities, see mathed.page/kinesthetics

# Part II
# Classroom Practice

# 4

# Cooperative Learning

Cooperative learning, also known as group work, is an approach based on the idea that students learn better if they have a chance to discuss their ideas with their peers. We do not claim that this absolves the teacher from playing a leading role in the classroom. Quite the opposite: effective collaboration among students requires serious involvement from the teacher as we will show in this chapter.

An early argument for group work was articulated by Chakerian, Stein, and Crabill, in the teacher notes to their groundbreaking 1970 textbook, *Geometry: A Guided Inquiry* (which we already mentioned in Chapter 2).

> In this course, students collaborate in solving problems and working exercises during class. The small-group learning method encourages participation by all students, promotes the exchange of ideas, and lessens frustration with difficult problems. Students are involved in the teaching and learning process and take more responsibility for their own learning. They actually discuss mathematics with their classmates. Teacher-led presentations and discussions are interspersed among small-group activities as needed.
>
> The cooperative learning approach facilitates mathematical exploration and problem solving. It encourages creative thinking, risk taking, critical thinking, and the sharing of ideas, all of which are important factors in mathematical problem solving.

Group work is especially helpful in managing a large class: instead of dealing with, say, 30 students, the teacher is dealing with eight to ten groups.

While in our view student collaboration should be the default in the classroom, it does not mean that every task must be done collaboratively. In fact, most of the time, students are working individually, but they know that if they get stuck they can get help from a neighbor. They also know that they are expected to provide such help when it is requested, and that in fact, that will help deepen their own understanding. Students can compare their answers as they work, which may alert them to possible mistakes, and suggest the need for actual collaboration. In cases where many calculations must be done before drawing any conclusions, that work can be shared.

Some tasks, to be sure, are best done alone – for example mechanical drills, or straightforward applications of learned procedures. More generally, group work is not appropriate or helpful in a classroom where those tasks dominate.

## Group Work: Why

Teachers who try cooperative learning techniques once and decide "it doesn't work" underestimate the needed cultural shifts, and the immense benefits their students would reap if these shifts happen. Specifically:

- Students talk about math. They're the ones who need to! It is difficult to learn challenging ideas in silence. We know that when we're struggling with a tough problem, we need to talk about it to clarify our thinking.
- Stronger students can help the others. This requires guidance from the teacher, especially at first, but it pays off in the long run.
- The teacher hears where the real questions are. In a traditional format where the teacher does most of the talking, it is easy to lull oneself into thinking that students understand everything we say. That is rarely the case, even if our explanations are crystal clear.
- Once students are accustomed to group work, the teacher is free to talk to one group or individual when that is needed, without the class degenerating into chaos.
- When a teacher works with a group, those students listen, as the teacher is much more likely to answer questions they have than when addressing the whole class.
- Students are active, not passive. Math is not a spectator sport: it can only be learned by doing math.
- Much learning takes place when going over homework. At home, students don't usually have the benefit of readily available help, so a reasonable amount of homework helps them separate what they know and understand from what they are still struggling with. This makes going over homework in groups especially efficient.
- Students can see each other's work, which allows them to learn about clear writing, organization, and documentation of answers.
- Students start to understand that they are in charge of their own learning, and that it does not all depend on clear teacher explanations.
- Students will practice articulation of mathematical ideas, argumentation, and evaluation of their peers' arguments.

- A student who may be reluctant to share an idea with the whole class is more likely to share it with a group of peers. From there, the idea can be shared more broadly.

On the other hand, group work has some disadvantages:

- Keeping track of students' work is a little harder.
- There is some loss of accountability in homework. This can be remedied by doing a quick daily survey of who's done it, by assigning occasional problem sets that get turned in, and by periodically quizzing students on new material.
- Some students will try to get a "free ride". It is essential to make clear up front that most tests and other assessments will be individual, and that the purpose of the group work, paradoxically, is for individual students to learn.

Sometimes students or their parents may complain that group work is holding them back. This concern is best addressed by making sure the course is challenging, every day.

## Group Work: How

Simon Fraser University professor Peter Liljedahl's research on "the thinking classroom" supports *visibly random* groups, reshuffled daily, working on vertical, erasable surfaces (in other words, whiteboards.) He writes:

> …the use of visibly random grouping strategies, along with ubiquitous group work, can lead to: (1) students becoming agreeable to work in any group they are placed in, (2) the elimination of social barriers within the classroom, (3) an increase in the mobility of knowledge between students, (4) a decrease in reliance on the teacher for answers, (5) an increase in the reliance on co-constructed intra- and inter-group answers, and (6) an increase in both enthusiasm for mathematics class and engagement in mathematics tasks.[1]

Group participants working on vertical whiteboards tend to be quicker to start, bolder in generating ideas, and more eager to discuss them. This may be due to the fact that standing reduces anonymity and that the ability to erase reduces the feeling of risk associated with writing down an idea.

**Our Recommendations**

Naturally, Liljedahl's techniques are not the only possible approach to group work.[2] We now share ours, beginning with some overall tenets, and then moving on to advice for specific situations.

Tenet #1 is to randomize and change groups regularly. It seems that groups of three or four work best (two or five if necessary). Changing the groups often enough allows various combinations to turn up. It makes no assumptions about kids and, because it changes regularly, no one can be too unhappy. Almost any outcome of the random selection has its advantages: for example, if stronger students end up together, that group will be more self-sufficient, thus freeing the teacher to concentrate on those who need more help. Finally, the random generation of groups relieves students from possibly toxic trains of thought such as "Am I supposed to be the dumb one in this group?"

Students should be aware they are responsible to help each other, irrespective of who is "ahead". However, unlike other proponents of group work, we do not ask students to wait for each other. We discourage this because it would set the pace according to the slowest students more than is appropriate. If some students are working much faster than others, you can and should include especially challenging ("sponge") problems at the end of each day's assignment. Those would be "semi-optional" in that you would not expect every student to complete them, but you would encourage the speed demons to take them on.

This is delicate, as you do not want to imply that other students *could not* handle these problems – the message should be: given time constraints, our goal is for all students to master the key ideas in the lesson, not for all students to do every single problem on the page. Note that this approach keeps the faster students in the classroom community, still available to help their group. This would not be the case if they were assigned a completely unrelated challenge problem.

As students work in groups, the teacher's role is to:

- *Be active at all times*, circulating among the groups, answering questions, coaching, and asking questions.
- Develop peripheral hearing and vision: while working with one group, you still need to keep the other ones on task.
- Encourage active listening.
- Help students learn how to teach, not tell.
- Take each group to their next step, rather than aiming for a least common denominator for the whole class. Again, aiming high is to everyone's benefit, as it keeps strong students engaged, and gives the others something to strive for.
- When appropriate (at the very least once per period), interrupt group work and lead a whole-class discussion. If the work is so easy that this is not required, it is probably too easy.

If the groups are working well, you may be tempted to use the opportunity to help an individual student who really needs that attention. Some proponents of group work would object to this, but we encourage you to trust your intuition, taking into account the whole picture: can this student be helped by a peer? Can you facilitate this? If not, and you choose to help this individual, it is essential to remain aware of what else is going on in the class.

With these guidelines in mind, we move on to specific advice for various segments of the operation. You will see that we prefer an informal collegial setup, with extremely hands-on teacher involvement. Unlike other proponents of group work, we do not assign specific roles within the groups (facilitator, recorder, presenter…). We'd rather have all the students' energy focused on the math.

**Doing the Rounds**

When you first assign a problem to work on in small groups, there may not be much for you to do. No one is stuck yet; no one needs your help. There is a lot you can accomplish in this time. You will want to quickly "do the rounds". Visit each group once just to check that they have gotten down to work. In the first round, look for any trouble with the wording of the problem that may be holding people up. If it's part of their job to decipher it, encourage them to do so. If it's a mistake, or if you need to supply a definition, then make a quick announcement. Ensure that students know what they are expected to do the minute group work begins. This might be reading through a problem, or there may be an instruction to copy some text, create a manipulative by folding paper, etc. One of the most effective devices is to be omnipresent. Any group not on task will find you standing next to them, watching. In many cases, this is all it takes.

If it's not just isolated groups, but it seems the whole class is getting away from you, then you should examine what you're doing that promotes doing something other than the math. You could be leaving some groups working too long, with nothing to do while other groups finish. You could be joking with them too much during class time.

On your second round you can linger longer. This is a good time to make a mental note of which groups are going faster than the others. It helps, during the subsequent large group discussion, to have a good idea of who has gotten how far. If a group has quickly and incorrectly or incompletely answered a problem and gone on to another, this is a good time to ask (innocently) for one of them to summarize for you what they found. The correct question on your part can cause them to re-examine what they've done without feeling that you've invalidated their answer (point of philosophy: you want them to be able to criticize their own work, realizing that mathematics will determine whether they are right, and that what they discover about this cannot be overruled by the teacher).

Robin writes:

For example, once I observed a group who had quickly concluded that there were no solutions, based on the incorrect manipulation

$$(a+b)^2 = a^2 + b^2$$

I chose to hand them a solution and have them figure out how to resolve the contradiction. I did not settle for "Gee, Teacher, your solution works so I guess we did something wrong." Instead, I insisted that they were not done until they were able to pinpoint what went wrong.

The most important outcome here is to instill the belief that it all makes sense, and that even though they are the students and not the teacher, they have the capacity to retrace the work and resolve the discrepancy on their own.

The early visits are often no more than spot checks. Subsequent visits to a group can turn up a number of situations requiring intervention. Here is some advice about common knotty scenarios.

**"Help, We're Stuck"**

The first thing to check is whether it is really true that *all* members of the group are stuck. If that is the case, you need to intervene.

The twin dangers here are that some students will say they're stuck so you do the work for them, while students who are truly stuck will lose morale and waste time if they have to sit idly during class. Asking if they have any ideas on what they might try will prove embarrassing if the answer is no. Sometimes we do this anyway. Sometimes we replace the problem with a smaller one: "If you knew that $A = 15$ could you do the problem? Can you do the problem if you aren't required to make the number of cows and chickens the same?"

Sometimes we guess why they're stuck: "So the problem is you don't really know the definition of average speed?" An example of what might happen here is that they did know this but did not think of going back to definitions as a way to proceed. Now when they say no that's not the problem, we know the definition of average speed, we can say, "Oh, then you must be saying you don't have any way of determining the quantities defining speed, such as the time or the distance". If the student then says how they will proceed, you can smile and be on your way.

In other cases, a student can be stuck because they don't really understand what's being asked. You can ask the group to rephrase it or ask them how they would check if someone else's answer (here you specify it) was right. It helps to have snooped enough so you have a good guess of where they're stuck. If you don't, you can ask them but won't always get reliable information.

If you have previously addressed problem-solving tactics in this class, then you might suggest to the students that they get out their list of problem-solving tactics and try some. It will do the students more good if you ask them to try a technique than if you tell them more precisely what to do, so try to keep your list of techniques down to a few that almost always help. Doing a problem with numbers instead of variables, doing a problem with smaller numbers instead of the given numbers, or sheer trial and error, are almost always useful.

## Getting Students to Work Together

This aspect of small group work is very dependent on culture. Many schoolchildren love the interaction and figure out on their own how to work together without much of a problem. What if your kids are not like that?

You need not expect it all to come together in the first week, but there should be some clear expectations from the outset. It can be fine for students to diverge on their own paths; teachers need to make clear when something is a group responsibility. All discussions must be respectful. If someone asks you for help, you will need to decide whether to offer it or just redirect the question to a group member. There is no universally correct response in that situation. Of course, there are many times other than this when the teacher will offer advice, unsolicited.

You may at times need to tell a group explicitly "Aisha has found what she thinks is an answer but Steve and Brenda apparently don't understand what she did, so Aisha, you're going to have to explain it and see if you can convince Steve and Brenda". If you want to try other things first, before being this explicit, ask Brenda what her group has found so far, and don't let anyone else answer for her. If she says she's stuck, ask if her whole group is stuck, and if not, tell her you'll come back in 3 minutes and ask her again for a summary of what her group has done. Make sure groups are sitting in a tight circle, not a line or a disarrayed cluster. Students should be discussing with each other face to face, not face to side of head.

Particularly with older students, one might say: "I just can't work with Julia - she won't listen and hogs the discussion". If you are reshuffling the groups frequently, as you should, it's probably fine just to tell the student not to worry, Julia will be moving on next week, or tomorrow.

## Common Obstacles

In some schools, collaboration in math class is a departure from the dominant culture. In all schools, there are students for whom this arrangement is challenging. We now discuss how to handle that reality.

**Free Riders**

There can be students who are content to let others do the work. Find a way to positively get them involved. Often, unwillingness to engage is a problem with confidence. With a student who needs some nurturing, it's probably better not to call them out on it but to make a note of it; then, in future visits to that student's group, pick that student as the focus for your questioning. For example, come back and, seemingly at random, ask that student to explain what their group has done. Do this repeatedly for a week, overweighting the one student in your sample but still asking the others sometimes. Involving the student as much as possible, with questions that are at their level but not patronizing, is often all that's needed. At the same time, impress upon all the students that it's everyone's job to make sure everyone is keeping up with the group. If the group will have a task such as presenting or writing some of the work, picking a less engaged student to be in charge can be a good choice. This motivates everyone else to make sure that at least their ideas are penetrating the other students' awareness.

**Students Who Are Behind**

By students who are behind, we don't mean in the one lesson, we mean the student has fallen behind on a longer-term scale, jeopardizing their ability to do the work.

First, it's a good idea to know what you can do for them and what you can't. Not every student will get to the same endpoint. A student who misses class a lot, or who started the year woefully unprepared, probably will not achieve mastery before it is time for the rest of the class to move on. Early identification can help, especially if there are resources for tutoring.

This being said, probably a group learning environment is one of the best realistic options for such a student. The student will be involved in discussions of ideas, will see other students modeling how to think through things, will have some de facto peer tutoring, and will be able to influence the pace at least a little. Meanwhile, you have to make sure they get the most out of their group work and don't drag down the group (they are as much afraid of this as you are). Here are the kinds of things you can do to help that don't require unrealistic commitments of your own time. One is to make sure the student who is behind does not drop out altogether. If Joe is struggling, spend some time around his group and be ready to pounce on those times he comes up with a good idea. To boost confidence, assign credit: if Diego figures out how to do something and it can be seen to be related to something Joe said, then it's "Joe and Diego's method".

Try to give Joe some extra constructive comments on his work (we try to do this for everyone, but when time does not permit, we concentrate on students who need it the most). Making sure Joe is keeping up with their group is delicate – asking Joe too many times to explain to you where his group is will cease to be productive. At that point, and in fact even better before that point, it will be necessary to find some way to offer Joe support outside of class.

## Students Who Are Ahead

Having such a student can be a real boon if they are gifted teachers as well. If they have a good feel for how to explain things and help others, they will make your class run more smoothly than you can on your own. Even in this case, avoid treating them in front of the class as a reliable source for the right answers. You don't want to create a situation where calling on them is tantamount to telling the class something yourself. It is OK though, to treat them as a reliable source for intelligent commentary.

Keep an eye on "Einstein" to make sure that this student is not explaining things to others before they have a chance to figure it out themselves. Let Einstein explain things at the board in situations where you know there will be some wrong or unclear stuff in the explanation. Make sure though, that you give Einstein as much encouragement for what was right and clear as you would another student. If Einstein is a loner and tends to work fast but not share with the others, it will probably work out fine. Sometimes, you can try asking Einstein explicitly to figure out a hint to give the rest of the group as to how to proceed but that won't completely solve the problem for them. It will make Einstein summon up teaching skills that are worthwhile in general, so it's worth a try, but be aware that he or she may not be capable of this.

## Some Other Obstacles

- Students who do not do homework. This is a serious problem if a substantial amount of class time is dedicated to going over the homework. It means that those students are wasting valuable instructional time, and not benefitting from one of the key advantages of this system. This issue must be addressed early and vigorously. It is usually less of an issue if homework is reasonably short, and lagged behind class work.
- Students who want to get ahead in order to have less homework. This is best addressed by separating class work from homework, for example by lagging the latter. (We discuss lagging homework in Chapter 9.)
- Students who are cynical and sarcastic. Students who hurt other students' feelings. It is most effective to talk to these students outside of class.

Most of the remaining obstacles are helped by gentle interventions in the group process by the teacher:

- Students who refuse to help others.
- Students who are very silent and rarely say anything.
- Students who sit too far so they can't be heard and their papers can't be seen.
- Students who just want to work with a certain person and no one else.

Ask for the behavior you want directly, rather than by making speeches. In other words, "Could you push your chair in and help Tom with #3?" rather than "It is nice to help others". Over time, this helps to build a collaborative classroom culture.

Before ending this chapter, we discuss one more challenge we face when students encounter a dead end when working in a group.

## Dead Ends

"I'm stuck!" This is the most common obstacle in inquiry-based teaching and yet the solutions can be tricky and elusive. What can a teacher do to get things moving, without undermining the inquiry part of the lesson? There are many types of dead ends. Recognizing the difference between these is a big help in deciding what to do next.

### Flaming Dead Ends

Robin writes:

> Perhaps the best kind of dead end is a *flaming dead end*, a line of reasoning that crashes and burns unmistakably. A flaming dead end yells out, "Where you have arrived is undeniably wrong; how did we get here?" A flaming dead end results from an incorrect line of reasoning leading to a consequence so patently false that the students are compelled to re-examine the road that got them there. If you see your students headed for one of these, then all you really need to do is encourage them to get there without undue delay.

For example, suppose you are using this question to introduce a new topic in probability: For the Eagles to make the playoffs, they either need to beat the reigning champions tomorrow (30% chance) or have Dallas lose tomorrow (20% chance). What is their chance of making the playoffs?

A good number of adults will add the two chances and get 50%. The right answer is not obvious; the first step is to see that 50% is wrong. You can tell them why, but the flaming dead end is a more powerful lesson. What if instead chance the Eagles win tomorrow is 60% and the chance Dallas loses tomorrow is 50%. Great, the chance the Eagles make the playoffs is 110%! Once students are forced to confront the fact that these don't simply add, they are in a much better position to work out what really happens.

A lower grade-level version of more or less the same concept arises in this story which could perhaps arise in a unit on Venn diagrams, or logic, or counting.

> It's raining on the day of the field trip. Luckily 15 students brought umbrellas and 8 have waterproof ponchos. How many will be able to go without getting soaked?

If the students agree on an answer of 23, you could mention there are only 20 students in the class. This becomes a low-threshold high-ceiling problem for youngsters. A discussion may unfold like this: (1) Explain what went wrong; (2) Try to figure out the right answer, maybe using a picture; (3) The answer is not determined, you need to know how many have both. (4) Answer it if six students have both. (5) Can you formulate a general answer?

Another example, from a higher grade where students have learned the meaning of the term "average speed": on a round trip between two points 30 miles apart, a car goes 30 MPH on the outbound journey and 60 MPH on the way back. A group computes the average speed as 45 MPH and is then stuck on part 2 of the problem because of their mistake on part 1. This is a natural mistake, one that would made by many adults, so you can expect this to happen and be prepared for it. A way to make this dead-end flame out is to change one of the two speeds to an extreme value, either very slow or very fast. What if the return speed is a million MPH. The return trip is essentially instant. Does that make the average for the round trip 500,015 MPH? Referring back to the definition, students can see without actually computing that this is not possible for a trip that took over an hour and remained on this planet. Or make the return speed zero. The car breaks down in Grandma's driveway and never makes it home. The round trip does not happen with a 15 MPH average speed.

It's possible, even likely, that the students will see the mistake but don't know how to right the ship. Try asking them what definition of average speed they are using. If they're using the wrong one, this should help. If they're using the right one, a few follow-up questions about the computation should do the trick. Students who have gone through this will almost certainly absorb the intuition and understanding better than those who are told, "You did part 1 wrong. Please try again with this formula".

Intuition is not the only benefit. Learning to trace back a logical error is developing argumentation, a crucial habit of mind which will help with reasoning that can be applied to almost anything: management, medicine, building construction, etc. Telling students where they went wrong might more efficiently get them to answer this problem correctly, but won't do them much good on any problems that are not clones of this one. In all these examples, the flaming dead end does not solve the problem. Rather, it goads students into rooting out a logical flaw that was preventing them from getting a reasonable answer. In so doing, it provides both motivation to work further and a mental model on which to build the next step.

Not all mistakes are this interesting. Many times, a dead end arising from a mistake can be corrected by saying, "Look! You dropped a factor of 2 right here". The more interesting the mistake is, the more value there is to grappling with and slaying the source of the error, and the more you should be inclined to push the student toward a flaming dead end.

**Zero Progress**

Some people think of a wrong answer as worse than no answer. This view is reinforced by the scoring rules on some multiple-choice tests. In the learning business, it is the opposite. Helping a student who has no answer is much harder than helping a student who has a wrong answer. Luckily, this type of dead end can be planned for. When you give students a problem to work on, part of the plan should be what you will do if they are stuck right off the bat.

This is a longer story than we can adequately cover here, but we can give an example of a kind of intervention that often works. Suppose a group is trying to compute the intersection of two lines. They say they have no idea how to start. Try telling them the answer is the point (3,5), or any point it's obviously not. When they say it's not, you ask them to justify that response. This demonstrates to them that they did actually know something about the problem. It also forces them to name a test that (3,5) fails, which means they know a test that the right point will pass. If this does not nudge them to the necessary algebra, you can probably figure out what to do next. You say, "Oh you're right, it's not (3,5) it's (3,−2) (where this is chosen to lie on one of the lines but obviously not the other). And if that does not work? Well, at least you now know that some group of students has not absorbed anything you thought they did in the last month! They're probably not the only ones. You will need to ascertain how widespread a problem this is. You might need to teach the whole unit again with a different approach.

## In between

In between zero progress and a flaming dead end are dead ends where a student has made some progress, has not reached any conclusions right or wrong, and does not know what to do next. These are harder to plan for because you don't know in advance what each student will have done before grinding to a halt. When we say we learn something every time we observe another's class or are observed ourselves, it is this kind of judgment that we are talking about. What is the right question or hint for a given situation? The answer lies in a combination of experience, books such as NCTM's *Principles to Actions*[3] and *Five Practices*,[4] and attention to the goals of the lesson and of long-term understanding.

You can try giving a hint. Students sometimes use "we're stuck" to game the teacher into giving a hint when one is not needed. This is less of a danger if you can see they did some work and then got bogged down. This is especially effective at an advanced level where the group does know a way to proceed but it would be dauntingly messy. You can tell them yes, I think your idea might work but it would take forever, so try this instead. It turns "stuck" into "basically you got it but I'll save you some work".

You can try going back to the basic list of hints we gave at the end of Chapter 2. Number 1 is: try a special case. When algebra is involved, this often means trying it with a number instead, or with one fewer term in the equation or one fewer variable. Could you recognize a solution if you saw it? Can you restate what you are trying to find? Did you use everything you were given? If you had to take a guess, what, roughly, do you think the answer might look like? A small number of such prompts, matched correctly to the situation, gets you through most of these situations. It moves the needle off of zero on the speedometer of progress and it boosts your students' confidence that they will get there in the end as drivers, not passengers.

## Individual Growth

The essential purpose of group work is the intellectual growth of individual group members. The group is the way forward, not the destination. This needs to be clear to the teacher, who should not merely monitor the groups' overall functioning, but also learn to spot specific students who are not getting enough from the arrangement and figure out what interventions are called for.

This essential purpose also needs to be made clear to the students: "Our goal is for each one of you to be able to do this work on your own. You work in groups, but you are in charge of your learning! You'll get a check on your understanding when you do homework and take tests, but you'll be better prepared for those if you make good use of group work".

## Discussion Questions

1. If you have not used group work, what are your reservations about it? If you have, what are the challenges and benefits?
2. We suggest collaborative work as a daily default. Do you agree? Are there certain types of activities that students might benefit from tackling in isolation? Which ones, and why?
3. Students cannot rely solely on each other to learn math. How does the teacher maintain their essential leadership role in a collaborative classroom?

## Notes

1. P. Liljedahl. The affordances of using visibly random groups in a mathematics classroom. In *Transforming Mathematics Instruction: Multiple approaches and Practices*, 127–144. Springer, New York, 06 2014. More recently, Liljedahl wrote *Building Thinking Classrooms in Mathematics* (Corwin Press, 2021). In that book, he shares the results of 15 years of research on how to get more students to spend more time thinking in more classrooms.
2. Another strategy for effective group work, known as *complex instruction*, was originally developed by Elizabeth Cohen, a professor at Stanford's School of Education. See her book with Rachel Logan: *Designing Groupwork* (3rd edition, Teachers' College Press, 2014) and *Mathematics for Equity: A Framework for Successful Practice*, edited by Nasir, Cabana, Shreve, Woodbury, and Louie (NCTM, 2014). Complex instruction has a solid track record in improving students' access to mathematical thinking. Henri has seen it in action in a wide range of classrooms, and was impressed by what he saw.
3. National Council of Teachers of Mathematics NCTM: *Principles to Actions*. Reston, Virginia, 2014.
4. M. Smith and M. Stein. *Five Practices for Orchestrating Productive mathematics discussions*. NCTM, Reston, Virginia, 2011.

# 5
# Learning Tools

The Common Core State Standards (and thus curriculum standards in many states) include this Standard for Mathematical Practice: "Use appropriate tools strategically". We wholeheartedly support this.

This chapter and the next two are about learning tools, their uses, and misuses. By tools, we mean a broad array of pencil-paper, manipulative, and electronic tools that can facilitate instruction. We include *conceptual tools* such as the Venn Diagram, which might also be called representations. We include *manipulatives*, these being objects that students move and rearrange to represent mathematical ideas. We include *computation engines* such as calculators and spreadsheets that provide computing power. Some tools of course occupy more than one category. Spreadsheets can be end-user tools or computation engines. Slide rules can be computation engines or manipulatives.

The calculator is the subject of much controversy, but it is only the tip of the iceberg. As with calculators, most instructional tools can be used in productive or unproductive ways. Keeping in mind what we can expect to gain from each kind of tool allows us to build effective lessons. We start with a reasoned discussion of the benefits – cognitive and attitudinal – of various kinds of tools.

## Why Tools?

There are several types of benefits that tools bring to the classroom.

One type is attitudinal. We want students to be active, engaged, unafraid, and collaborative. Tools give students something to interact with. To the extent that the activity seems meaningful, students have a greater chance of being engaged when involved in it than when listening. To the extent that the activity allows students to make choices, they experience self-determination, and that creates ownership of the resulting knowledge. Some tools are particularly good at creating a circumscribed environment. This helps to build confidence and initiative; the universe of possibilities is finite rather than dauntingly large; students can navigate on their own and achieve mastery.

Secondly, most tools provide cognitive benefits. Much mathematical learning involves making connections and mapping knowledge from one domain to another. The concept of a function is important but it resides at a level of abstraction that challenges many students. A function can be an equation, a rule, a set of inputs and outputs, or a graph. Making connections between these is a big part of understanding functions. Tools can help to make these connections more obvious, more familiar, more automatic, and more concrete.

Indeed, we often confuse familiarity with one representation of a mathematical concept with a full understanding of the concept. In fact, the ability to look at the same idea in different ways is a crucial ingredient of understanding. For example, when a student seems stuck, it rarely helps to re-explain the same idea in the same way. If as teachers, we can come at it from a different angle, it's a sign that we have a deeper, multifaceted understanding of the concept. Tools are one way to get at this.

The conceptual tools we discuss in this chapter are such representations, positioned strategically between the others, or adapted to be more accessible. Manipulatives, on the other hand, are actual physical objects. These, according to Seymour Papert,[1] serve even more basic cognitive functions. In this passage, he discusses how a simple object, a gear, serves as a conduit. Mathematical concepts hitchhike into the brain on the backs of existing motor and sensory knowledge.

> What an individual can learn, and how he learns it, depends on what models he has available. The gear can be used to illustrate many powerful "advanced" mathematical ideas, such as groups or relative motion. But it does more than this. As well as connecting with the formal knowledge of mathematics, it also connects with the "body knowledge" of a child. It is this double relationship – both abstract and sensory – that gives the gear the power to carry powerful mathematics into the mind.

Some see Papert's argument about *objects-to-think-with* as confirming the need to teach to different learning styles, in this case, "physical learners". Whether learning styles actually exist is subject to debate, but we take the cognitive benefit provided by manipulatives to be more general. Whether or not some students learn best in particular channels, everyone responds to all of the primary channels for mathematics: sight, sound, and touch. Using all the channels is an effective practice from any viewpoint and for any student.

Finally, tools can serve a low-level but necessary purpose, namely to provide physical variation and novelty. Some tools, such as electronic math games, are popular because they are just plain fun. Many appeal visually to kids. Electronic games have an advantage over the old-fashioned kind because they somehow provide superior sensory gratification. Almost all tools involve physical motion on the part of the students.

To summarize, here are the central arguments for a tool-rich pedagogy:

- Paper-pencil, manipulative, and electronic tools have the potential to make the classroom a more student-centered, collaborative environment, where learning is more likely to take place.
- They can help generate the sort of discussion and reflection that is a prerequisite for deep understanding.
- They can make some challenging ideas accessible to more students, while at the same time offering an opportunity for all students to make connections they may otherwise miss.
- In particular, they can help bridge the visual and the symbolic, a connection that is fundamental to all mathematics and science.
- They can increase student self-reliance by providing a circumscribed world that they can navigate and master.
- And last but not least, they can bring variety to math classes, which are unfortunately often seen by teacher and students alike as boring but necessary drudgery.

All this adds up to an overall improvement in the classroom atmosphere. In addition to the cognitive benefits, it can make all the difference with respect to classroom management, engagement, and student enjoyment.

**Pitfalls**

At the same time, there are pitfalls.

Deborah Ball calls the first of these pitfalls "magical hopes". In a 1992 paper,[2] she challenges the belief that students will automatically come to the conclusions their teachers expect merely by using manipulatives such as base ten blocks. This is true even or perhaps especially if students "correctly" use manipulatives according to a teacher-provided procedure. There is no magical transmission of understanding from the fingers to the brain.

Similarly, some people believe that the availability of automated instruction on the Internet is a revolutionary development that will make math accessible to all. After all, students can watch a video as many times as they need to and then can test their understanding in a personalized test. This in fact makes teachers obsolete! Of course, it is not that simple. The wide availability of both free and commercial materials online has not yielded an explosion of success in math, nor can it.

In each case, the problem is lack of guidance. Lessons, tool-based or not, need to be constructed thoughtfully. When there is no tool, every explanation and every problem or question is planned by the teacher.

When using tools, it is easier for students to take initiative, which is good in some ways, as it develops their sense of agency. But it runs the risk of ineffective use of limited time. Low-tech manipulatives are no less susceptible to this than are electronic tools.

Manipulatives lessons in the style of "this is how you do it, now practice" work poorly because they lose sight of the goal. Electronic tools can be used in a similar "do this, do that, what do you notice?" format which suffers from the misconception that students will spontaneously notice the thing we want them to notice. In reality, what they notice is determined by what they already know and understand.

At the least, tools can distract us from the task of planning. Beyond that, many tools place high cognitive demands on both teacher and students. Spreadsheets, for example, are popular these days in high school math lessons. There is considerable overhead, though. Students need instruction in how to use them. Teachers may themselves need some instruction. There is always the chance the technology may fail, leading to hours spent watching IT support try to get the license key properly input, the screen projected, or yesterday's data to load. All of these subtract time from math instruction.

Lastly, with tools that are ends in themselves, there are questions about how much time should be diverted from math instruction in order to teach a platform. Proficiency in using something such as a spreadsheet, statistical functions in a software package, or a beginning coding platform are legitimate educational goals. To the extent that they compete with time spent learning mathematics, decisions need to be made. One winning strategy is to pack the process of learning the platform with mathematics you already wanted to teach. (For example, turtle geometry provides an environment where students can learn basic programming concepts along with basic geometry concepts. Or Fibonacci numbers can provide content for a lesson on how to use a spreadsheet.)

## Which Tools?

There are lots of specialized kitchen tools: peelers, corers, slicers, pizza cutters, zesters, melon scoops, and so on. A knife is a flexible tool, which can be used to do many things in many situations. Math educator Shira Helft suggests, "when you can, use a knife". If you have a good melon-baller lesson, that is great, go for it! But in general, if we are spending time training teachers and students to use one thing, it should probably be the knife. She gives an example of a knife: the rectangle model of multiplication. (We call it the rectangle model, not the area model because for young children, it is an array of objects rather than a continuous area, but that's another conversation.) That is indeed a good example, as it spans arithmetic (Figure 5.1), algebra (Figure 5.2), and calculus (Figure 5.3).

**Figure 5.1** Multiplying two-digit numbers.

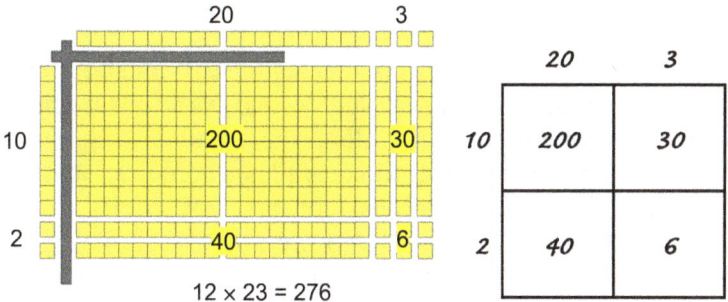

**Figure 5.2** Distributing and factoring.

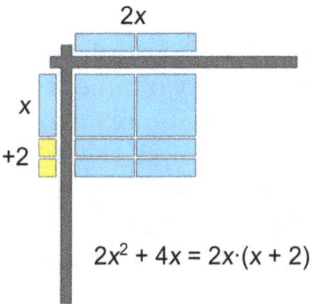

**Figure 5.3** The derivative of $y = x^2$.

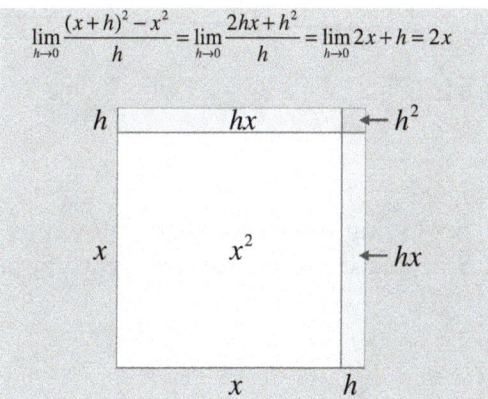

(See also "The Well-Chosen Rectangle" below for how the rectangle model applies to working with fractions.)

Naturally, there are other examples of multiuse tools, which will be explained when we return to them below: story tables, function diagrams, algebra manipulatives, the geoboard, electronic graphing (e.g., Desmos[3]), interactive geometry (e.g., GeoGebra[4]) .... They are tools not just for the student, but also for teachers and curriculum developers. By all means, prioritize the knives!

However, we should not limit ourselves to such a list. As one masters knife-level tools, there is nothing wrong with also adding specialized tools to one's toolkit as one gets further along one's career path. For example, one can use pattern blocks to introduce angles, geometric puzzles to illustrate scaling, and the circular geoboard for inscribed angles. There are also electronic tools which on the one hand are fairly specialized, but on the other hand, are so powerful they are definitely worth learning about. Here are two of those: Snap! for programming,[5] and CODAP for statistics and probability.[6] (We discuss manipulatives in Chapter 6 and electronic tools in Chapter 7.)

Actually, listing all these tools reminds us of a basic rule of professional growth: since you can't learn everything at once, prioritize! The only reason we have such a long list of tools in our repertoire is that we're old, and we have had plenty of time to learn them, and to develop activities for them. But anyone who plans on being a math teacher for a while should keep an open mind about learning new tools. Using more tools leads to a more varied classroom, more visual bridges to concepts, more student initiative and responsibility, multiple representations of the most important ideas, and a better way to preview and review material.

Our goal in the rest of this chapter and the next two is to catalog a few tools of each type, some of which are familiar and some undeservedly obscure.

## Conceptual Tools

### Exploding Dots

One such tool is "exploding dots", formerly known as "chip trading". It is a representation of place value. For example, Figure 5.4 shows the number 234:

**Figure 5.4** Exploding dots.

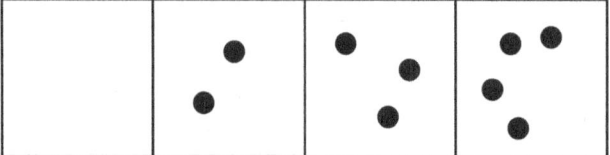

In this model, one dot can "explode" and be replaced by ten dots in the adjacent space to the right, or ten dots can implode and be replaced by a single dot in the adjacent space to the left. This makes it possible to model the operations of arithmetic and get insight into them.

In the chip trading model of the 1970s, we would *trade* a chip in one place for ten in the adjacent space to the right. James Tanton decided to call this trade an explosion, because (he told us) children like sound effects: "Kaboom!".[7]

**Number Lines**

Number lines are one way to represent numbers and operations. For example, a line can be drawn in the playground, and students can stand on 1, 2, 3, 4, and so on. The teacher can say "add 5 to your number", and students move accordingly, connecting arithmetic to motion. Alternatively, a clothesline can be taped to the wall, or strung across a classroom, and numbers can be hung at the appropriate places.[8]

**The Well-Chosen Rectangle**

Fractions, of course, can be represented in various ways. The *well-chosen rectangle* representation on grid paper facilitates computation and helps to introduce the idea of a common denominator.[9] This is an accessible model that can be the foundation of important discussions, and can be used for many different fraction topics. Of course, it should complement (not replace) other approaches.

Figure 5.5 shows an example: which is greater, $\frac{2}{3}$ or $\frac{4}{7}$?

**Figure 5.5** Comparing fractions.

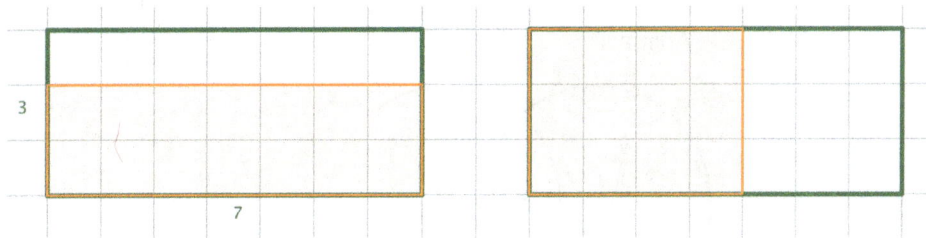

Using 3 by 7 rectangles (as suggested by the denominators) makes it easy to see both fractions as parts of the same unit. Counting reveals that $\frac{2}{3}$ consists of 14 squares, or $\frac{14}{21}$, while $\frac{4}{7}$ consists of 12 squares, or $\frac{12}{21}$.

Learning Tools ◆ 83

Which is greater has been made clear. The very same rectangles can help us think about $\frac{2}{3}+\frac{4}{7}, \frac{2}{3}-\frac{4}{7}, \frac{2}{3} \div \frac{4}{7}$, and of course common denominators.

You can use a single copy of the well-chosen rectangle to illustrate fraction multiplication. Figure 5.6 shows $\frac{2}{3} \times \frac{4}{7}$.

**Figure 5.6** Multiplying fractions.

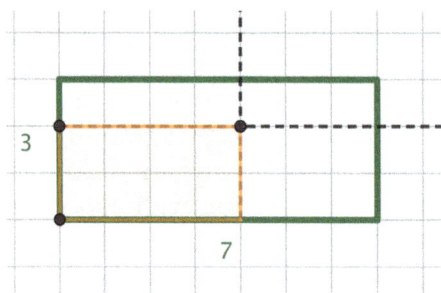

## Venn Diagrams

The Venn diagram is a common way to represent sets of items with partial overlap. Each set is represented by a circle: elements of the set are inside the circle and things not in the set are outside. To represent two sets simultaneously, draw two circles whose interiors overlap.

For example, one set could be animals, the other could be things in my house. Things in both sets such as a cat, are shown in the overlapping region, as in Figure 5.7.

**Figure 5.7** Venn diagram of things in my house and animals.

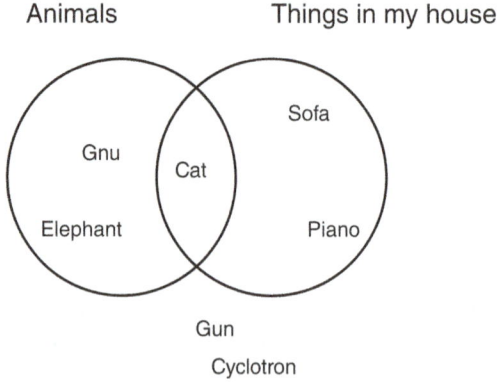

Things in only one set are shown within the corresponding circle but not in the overlap. There may be things such as a gun or a cyclotron that are shown outside of both circles because they are in neither set. When drawing many sets, if regions are required for all possible overlaps and no-overlaps, some of the circles have to be deformed as in Figure 5.8.

**Figure 5.8** Four sets and their intersections.

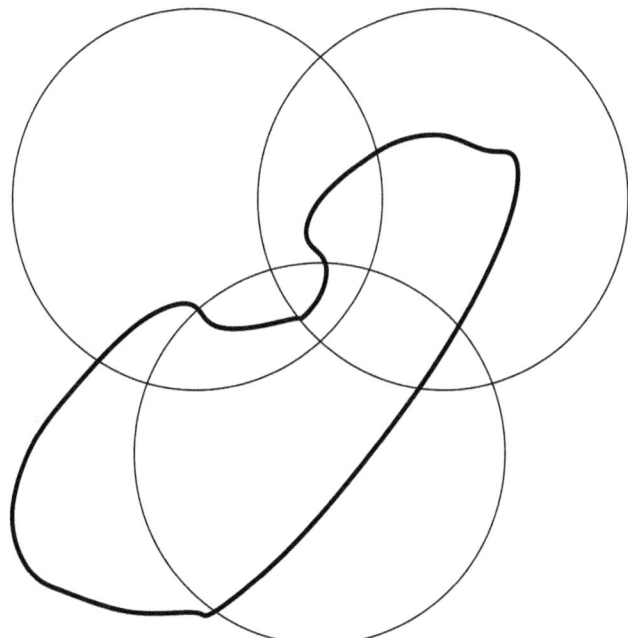

Sometimes one puts the numbers of elements in the region, instead of writing the names of all the elements. Many mathematicians and educators come to use Venn diagrams almost instinctively, as if they directly portray our mental image of a set. Despite the simplicity of Venn diagrams, however, students often need practice before they can use them correctly. The ice-cream problem, taken from a math competition for middle school students,[10] illustrates this.

> At San Leandro High School, 76 students like algebra, 63 like music, 57 like ice cream, 21 like algebra and music, 17 like algebra and ice cream, 13 like music and ice cream, and 5 actually enjoy all three. One student doesn't like any of these. How many students are there?

Many students initially write in numbers as in Figure 5.9.

**Figure 5.9** Oops!

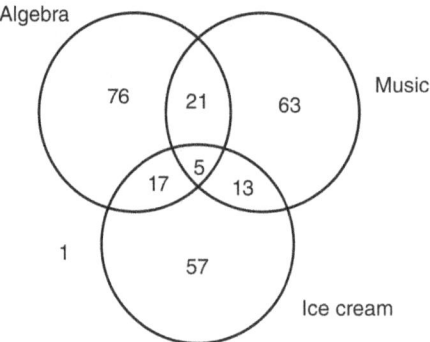

Only after some questioning do they realize what is wrong and how to fix it.

*Attribute Blocks*[11] are a manipulative tool well suited to introducing very young children to Venn diagrams. Each block is thick or thin; red, blue, or yellow; triangle, rectangle, square, hexagon, or circle. If one region is labeled "thin", and another is labeled "triangle", thin triangles will be in the intersection, and a thick square would be on the outside.

**Input-Output Tables**

Consider this table:

| $x$ | $y$ |
|---|---|
| 1 | −1 |
| 2 | 1 |
| 3 | 3 |
| 4 | 5 |

Can you see the pattern? If we were to extend the table, what would the next rows look like? What about the 100th row? What about the $n$th row?

In this example, the teacher gave an orderly set of inputs, making the rule a little easier to detect. Such tables can be the platform for a game of "What's My Rule?", where students provide the inputs (the $x$) and the teacher reveals the corresponding outputs (the $y$). They can be analyzed to provide information about the rate of change of a function, as the change in $x$ and the change in $y$ are easy to calculate. In particular, if the rate of change is constant, the displayed function is linear. In this case, $y$ increases in the same direction as $x$, but twice as fast, as shown in Figure 5.10.

**Figure 5.10** Graphing the points.

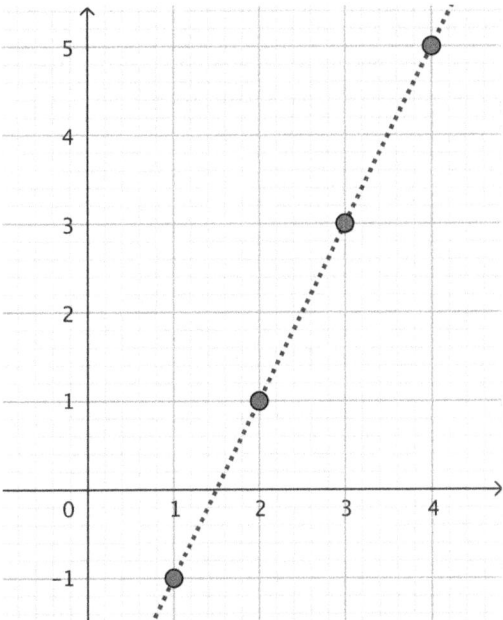

Other diagnostic tests can be performed on the data in a table.[12] For example, if adding a certain amount to the $x$ always leads to multiplying the $y$ by a certain number, that is characteristic of an exponential function, as in this example, where adding 1 to the $x$ doubles the $y$:

| $x$ | $y$ |
|---|---|
| −1 | 1.5 |
| 0 | 3 |
| 1 | 6 |
| 2 | 12 |

## Story Tables

Shira Helft points out that such tables, while useful in all the ways we suggest above, do not tell the whole story – only its beginning ("once upon a time", the value of $x$) and its end ("they lived happily ever after", the value of $y$). She suggests that much can be learned about equations and functions by using *story tables*, which apply order of operations to deconstruct the expression.

Here is an example for middle school or Algebra 1. Say we would like to solve $3(x - 10) = 15$. The table in Figure 5.11 tells the story: first we subtract 10 from $x$, then we multiply the result by 3. This provides a format

Learning Tools ♦ 87

for trial and error ("guess and check"). In Figure 5.12, we try 0 and 1 as values for $x$ and fill out the rest of the table accordingly. 0 and 1 did not work out too well, as we got −30 and −27 in the last column. (We were hoping for 15!) But at least we were going in the right direction! In Figure 5.13, we try a bigger jump: How about 10? This yields 0 in the last column, which is better, but not great.

**Figure 5.11** Story tables: the setup.

$$3(x - 10) = 15$$

| $x$ | $x - 10$ | $3(x - 10)$ |
|-----|----------|-------------|
|     |          |             |

**Figure 5.12** First two tries.

$$3(x - 10) = 15$$

$-10 \searrow \quad \cdot 3 \searrow$

| $x$ | $x - 10$ | $3(x - 10)$ |
|-----|----------|-------------|
| 0   | −10      | −30         |
| 1   | −9       | −27         |

**Figure 5.13** One more try!

$$3(x - 10) = 15$$

$-10 \searrow \quad \cdot 3 \searrow$

| $x$ | $x - 10$ | $3(x - 10)$ |
|-----|----------|-------------|
| 0   | −10      | −30         |
| 1   | −9       | −27         |
| 10  | 0        | 0           |

But wait! What if we put 15 in the last column, and work backward? To figure it out, we add arrows to the table, pointing from right to left, to show the inverse operations: ÷3 and +10. Figure 5.14 shows what happens.

**Figure 5.14** Working backward.

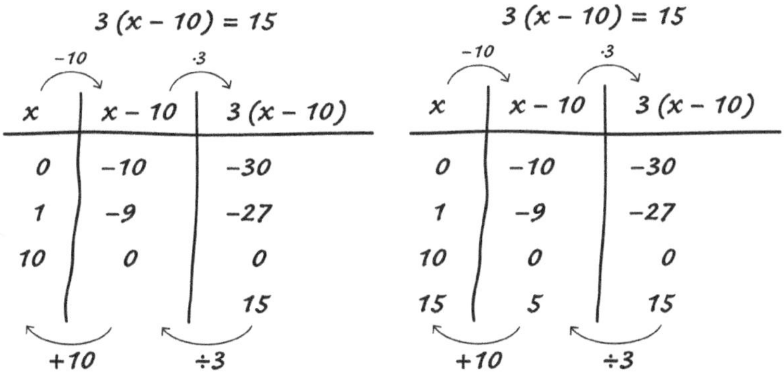

$$3(x - 10) = 15$$

| $x$ | $x - 10$ | $3(x - 10)$ |
|---|---|---|
| 0 | -10 | -30 |
| 1 | -9 | -27 |
| 10 | 0 | 0 |
|  |  | 15 |

$$3(x - 10) = 15$$

| $x$ | $x - 10$ | $3(x - 10)$ |
|---|---|---|
| 0 | -10 | -30 |
| 1 | -9 | -27 |
| 10 | 0 | 0 |
| 15 | 5 | 15 |

We got it! $x = 15$. Note that having gone forward first a few times is what makes this reverse use of the table readily accessible to students.

Story tables can also be applied to help develop insight into transformations of most of the functions students see in high school. Figure 5.15 shows an example involving absolute value.

**Figure 5.15** Analyzing a function.

| $x$ | $x - 4$ | $|x - 4|$ | $3|x - 4|$ | $3|x - 4| - 6$ |
|---|---|---|---|---|
| 2 | -2 | 2 | 6 | 0 |
| 3 | -1 | 1 | 3 | -3 |
| 4 | 0 | 0 | 0 | -6 |
| 5 | 1 | 1 | 3 | -3 |
| 6 | 2 | 2 | 6 | 0 |

Studying the table shows what operations result in moving and stretching the graph, and which one yields symmetry. For this to work, it is important for students to actually fill out the table: they probably would not appreciate all that's going on merely by looking at it.

More generally, a story table can be used in many ways, depending on which part of the table is revealed, and which part is left for the student to fill out. The examples given here only scratch the surface of this powerful representation.

### Function Diagrams

Function diagrams are a hybrid of the input-output table and the number line. In fact, they consist of two parallel number lines, with the output connected to the input with a straight line segment. Figure 5.16 shows an example for $y = 2x - 3$.

**Figure 5.16** Three representations of y = 2x − 3.

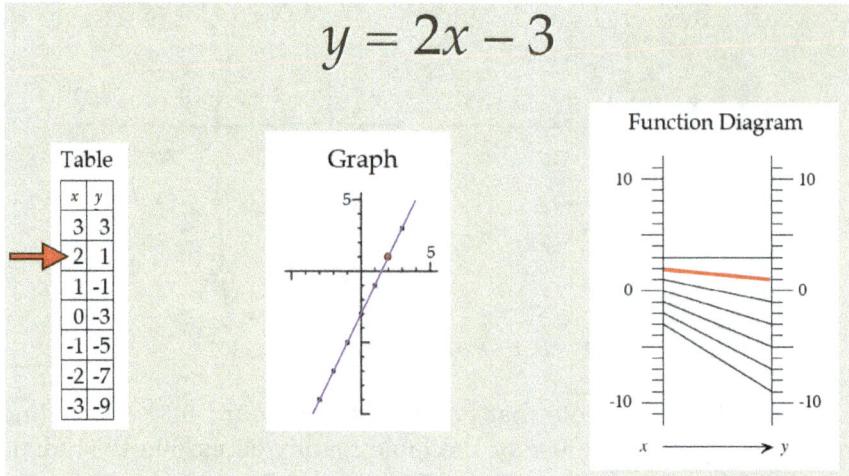

Function diagrams provide a context to discuss many mathematical ideas. For example, look at the pattern:

$$2 + 3 = 5$$
$$2 + 2 = 4$$
$$2 + 1 = 3$$
$$2 + 0 = 2$$

It can be represented by the function diagram in Figure 5.17. It is easy to figure out how the diagram can be extended, and thus find the value of 2 + (−1), 2 + (−2), etc.

**Figure 5.17** 2 + x.

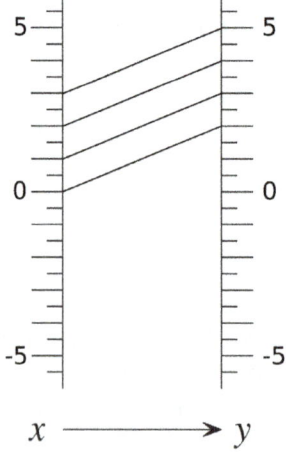

Or, consider the diagram in Figure 5.18, a representation of $y = \sqrt{x}$:

**Figure 5.18** The square root function.

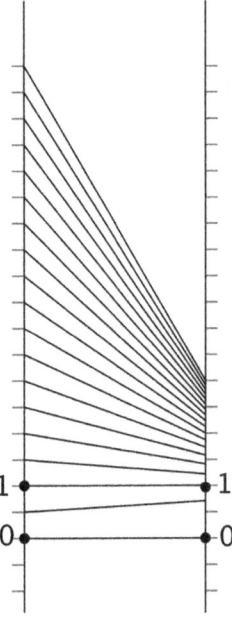

This illustrates many ideas: negative numbers do not have square roots in the real numbers; the square root of 0 and 1 are 0 and 1; the square root of a positive number less than 1 is greater than the number.

The goal, of course, is not to replace the Cartesian graph, but to complement it. While the Cartesian graph is a life tool, the function diagram is a learning tool – one that can be used to throw additional light on many important concepts. In particular, the composition of functions is much clearer than it would be in a Cartesian graph, as shown in Figure 5.19.

**Figure 5.19** Composition.

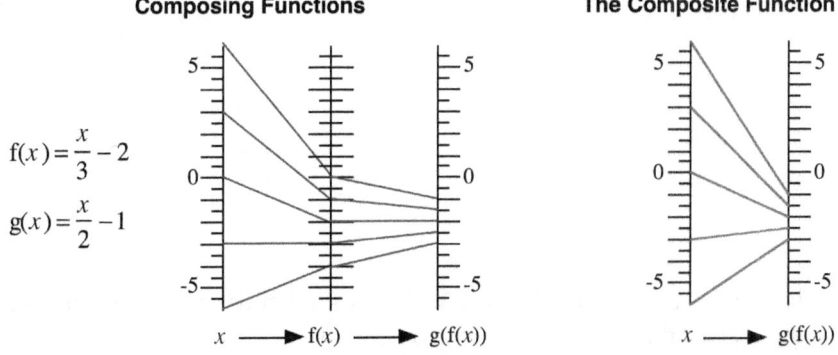

Learning Tools ◆ 91

This representation shows what happens to the rate of change when two functions are composed – a crucial concept in calculus.[13]

**The Ten-Centimeter Circle**

This ten-centimeter-radius circle, with centimeters and millimeters marked on the axes (as on a ruler), and degrees marked on the circle (as on a protractor), provides a way for students to find a reasonable approximation of the sine, cosine, and tangent of any angle by using a straightedge and the given markings. In Figure 5.20, you can see that $\tan(30°) \approx 0.58$.

**Figure 5.20** Using the 10-cm circle.

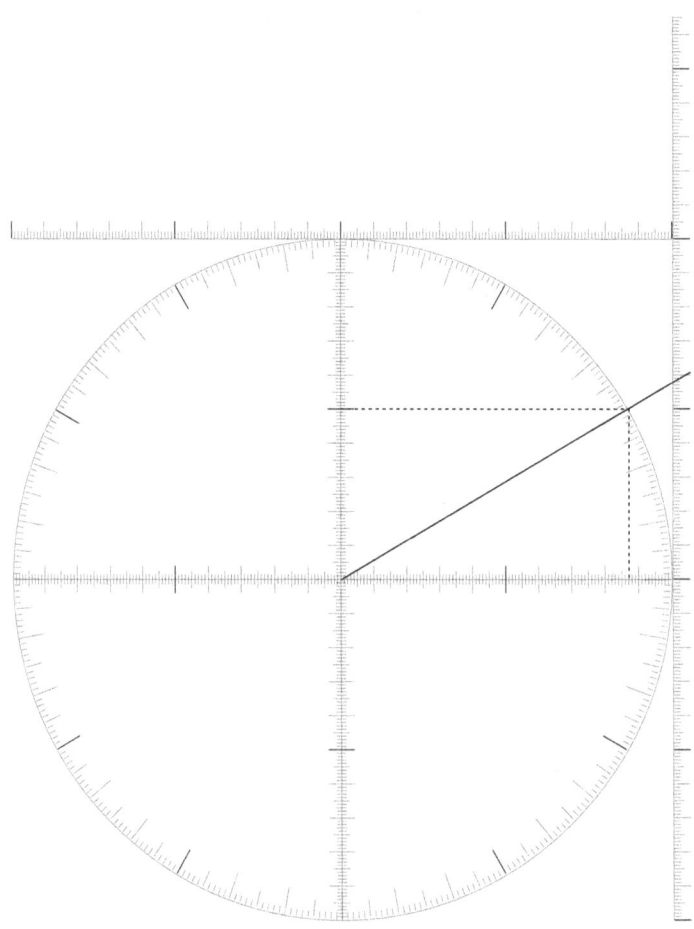

This tool can be used to solve problems without a calculator, such as "how tall is the flagpole?" (See Figure 5.21.)

**Figure 5.21** The flagpole and its shadow.

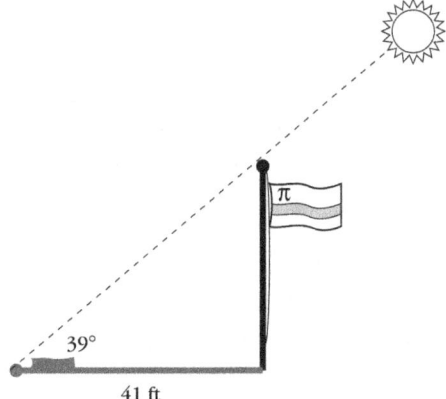

In fact, students can do this readily, without knowing this is trigonometry. The problems can be presented as a connection between angle measurement and slope. But this tool's most significant impact is that after using it for a while, students have a solid foundation for further work in trigonometry.[14]

## Multiple Representations

Inevitably, in our discussion of learning tools, we revealed the deepest argument for their use: they provide additional representations of mathematical concepts. As it turns out, this is not just good pedagogy: it is the nature of mathematics that the same structure underlies multiple representations. Does using multiple representations confuse the students? Well, yes, it may. It is especially confusing when connections are not made between the representations, or when the teacher's understanding of a representation is superficial. But we run greater risks if using a single approach. Multiple representations (and thus multiple approaches) are dictated by the math itself.

Here are some examples with whole numbers.

- A good way to think about whole number subtraction is with the help of counters. "If I start with ten counters, and remove three, how many are left?" Another good way is to think of whole numbers as sitting on a number line. "If I start on the 3 and count up to the 10, how many steps did I take?" Or, we can use Cuisenaire rods.
- For place value, we can use counters again, possibly using the "exploding dots" version, or an abacus. Or, as we will discuss

in Chapter 6, we can use base ten blocks, whether purchased or student-made.
- For multiplication, we can use counters again: 3 times 5 can be represented as 3 sets of 5 counters. Or, we can arrange the counters in a rectangular array, which allows us to see that $3 \times 5$ equals $5 \times 3$, and previews the area model for multiplication. Or we can do 3 hops of length 5 on a number line. Or we can learn to count by 5's – a condensed version of the number line model.

The power of math is precisely that the same structure (in this case whole number arithmetic) describes many different phenomena. Understanding these representations, and their relationships with each other is a lot more powerful than teaching this subject in a single way. It respects students' intelligence, and it is truer to the math: numbers are indeed all those things, whether you like it or not.

Is this confusing? If it is, it is because numbers are challenging, not because we approach them in multiple ways. If a teacher is comfortable with all the models, uses them strategically, and helps students think about the connections, multiple representations are illuminating because they reveal the many meanings of the underlying math.

The same is true about fractions: the widely used "pie" representation provides a great connection to angles and to time as seen on an analog clock. A number line representation emphasizes that a fraction is a number, and reveals its relative size. The "well-chosen rectangle" representation facilitates computation and helps to introduce the idea of common denominator.

Or take the use of letters in algebra. Sometimes, $x$ is an unknown, and to find its value we learn to manipulate symbols. Sometimes we use it to make a general statement about algebraic expressions, for example $2x = x + x$. In that case, $x$ can be any number. Algebra manipulatives can help with that. Sometimes it is a variable, and we see what happens to an expression as $x$ varies. Tables, graphs, and function diagrams can help us understand that. Sometimes $a$, $b$, and $c$ are parameters. They constrain an expression, but they are not unknowns or variables.

A well-intentioned mathematician once commented that variables are easy to teach: just patiently explain to students that they behave just like numbers! The high failure rate in the traditional Algebra 1 course is largely caused by the naive belief that patient explanations of how to move $x$'s and $y$'s on a piece of paper is the only way to go, as long as you follow it by mind-numbing and meaning-free practice. The reality is that it is a good idea to use many representations: tables of values, graphs, manipulatives, symbol manipulation, and function diagrams. Done well, this is not confusing: it's illuminating.

Or take trigonometry: the sine can be seen in a right triangle, or in a general triangle, or in the unit circle, or in a Cartesian graph. This is not because a teacher decided that. It's in the math itself – and it gives us multiple approaches to the concept, with opportunities to make connections between them.[15]

One last example: if in Algebra 2 complex numbers are introduced visually as points in the plane, with their own arithmetic rules, they are far easier for students to wrap their head around. This requires showing that those rules are consistent with real number arithmetic. A formal proof can be introduced in a post-Algebra 2 course.[16]

In short, multiple representations reveal the many meanings of math concepts. To make that less confusing, the answer is not to limit students to a single representation, even if you're proud of how good you are at teaching that particular one. Instead, give some thought to a proper sequencing of the representations, encourage students to use the representations that make sense to them, and especially be sure to connect the representations to each other.

### Seeing Is Believing?

We do not want to oversell visual representations. Indeed, we need to address a common misconception. Some educators believe "if they can see it, they'll understand it". This is a manifestation of one of the most harmful fallacies in math education, the illusion that students see the same thing as we do when they look at the same representation. In reality, what one sees is greatly influenced by what one already knows and understands. Visual representations do not automatically confer understanding.

We'll use *proofs without words* to discuss this. Those are proofs based on a visual representation of a theorem that provides a convincing argument about its validity without the need for any accompanying text. The genre has been much enriched by the increased availability of computer animation. This is of course relevant to math education: many of the concepts we teach can be illustrated visually, including with clever and sometimes beautiful animations. What is the proper place for such visuals in teaching?

First of all, we'd like to challenge the idea that those proofs should actually be "without words" in an educational context. In reality, proofs without words are mostly meaningful to people who already understand the result that is being illustrated. If we want them to support student learning, it is mostly through discussion, reflection, and writing. Our job as teachers is to help students come up with the words that explain these images.

For example, Figure 5.22 illustrates the "difference of squares" identity.

$$a^2 - b^2 = (a+b)(a-b)$$

**Figure 5.22** Difference of squares.

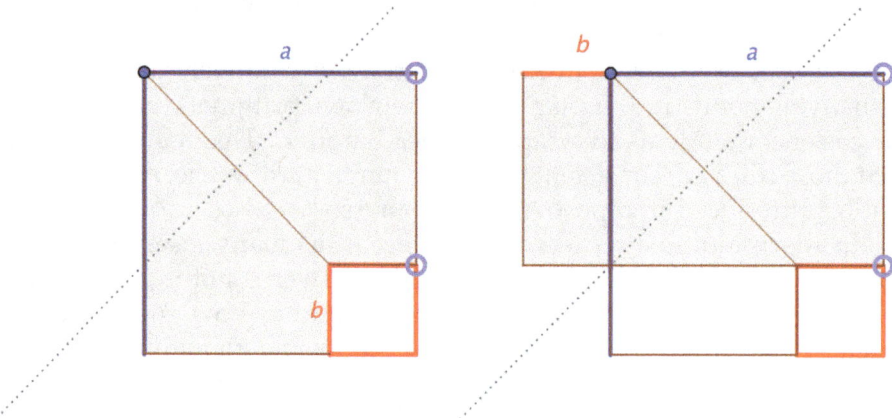

The point is clearer if you watch an animation of this.[17] Clearer for you, that is. For students, it only becomes clear if they think about, discuss, and/or write about the animation. The fact that we (math teachers) appreciate the elegance of the visual representation in no way guarantees that merely looking at it is helpful to students.

This is not at all to deny the important role visual representations can play in teaching. Rather, it is an argument to think case by case about where those fit in our lesson plans. One thing we learned about visuals in our many years of using them with students is that working with physical materials before watching the animation or looking at the figure helps prepare students to get the most out of the visual representation.

For example, the above illustration of the difference of squares could be preceded by a paper-scissors activity using grid paper. "Draw a square, different from your neighbors'. Draw a smaller square in one of its corners. Cut out the smaller square, and you will be left with the difference of the two squares. Make a single cut in the resulting figure, and rearrange the two pieces to make a rectangle". Doing this will only take a few minutes, and it just about guarantees that students will have a better understanding of what is happening in the animation. The discussion should happen first about the paper model, and then about the computer version.

Here is another example: There are many, many online visual proofs of the Pythagorean theorem.[18] Again, we suggest that you precede them with work using concrete materials (the geoboard, dot paper, grid paper, or puzzle pieces), and follow them with discussion and writing.

## Active Involvement

Just to be clear, we do not suggest the use of concrete materials, or any particular tool, because of a naive belief that those activities automatically confer understanding. Quite the opposite! The point is that actively doing something provides a better platform for discussion, reflection, and writing than watching something. There is a time for watching, but it comes later, and writing about what they see at that point provides a good way for students to assess how well they understand the concept.

For most students, learning does not happen merely by listening to teacher explanations, even if they are accompanied by beautiful images or clever animations. There are no shortcuts: if something is important, it is best to engage students intellectually in multiple activities that lead to the necessary reflection, discussion, and writing. Once that foundation is built, they can appreciate "proofs without words", and put words to them. More generally, active involvement with learning tools, plus discussion, helps students understand teacher explanations of the relevant concepts.

We discuss more tools – manipulative and electronic – in the next two chapters.

## Discussion Questions

1. Which of the tools discussed in this chapter have you used? How did it go? Which among the tools you have not used are you interested in learning more about?
2. For important concepts, we recommend using multiple representations, multiple tools, and multiple approaches. Some teachers worry that this would be confusing to students and that it's better to pick the best strategy and put all your energy into getting it across. Discuss.
3. Why is it important to use visual representations of mathematical concepts?
4. We discussed Venn diagrams in two different contexts, but we did not specify the benefits of using this representational tool. What are those?

## Notes

1. S. Papert. *Mindstorms*. Harvester Press, Basic Books, New York, 1980, 1993.
2. D. L. Ball. Magical hopes: manipulatives and the reform of math education. *American Educator*, 16:14–18, 46–47, 1992.
3. desmos.com
4. geogebra.org
5. snap.berkeley.edu
6. codap.concord.org

7. Tanton expanded the use of this model to polynomial arithmetic, and more. See globalmathproject.org
8. See also a kinesthetic number line activity to introduce complex numbers in high school: mathed.page/kinesthetics/complex.html
9. For a full introduction to this representation's many uses, see mathed.page/early-math/fractions
10. M. Sloper, A. Gulimovskiy, R. Pemantle, J. Wolinsky, and D. Bach. *Mathemania*. Black Pine Curriculum Institute, Berkeley, California, 2007.
11. Attribute blocks are available from many educational publishers.
12. For many ideas about recognizing functions from their tables, see mathed.page/recognizing.html
13. For much more on the uses of the function diagram at all secondary school levels, see mathed.page/func-diag
14. For more information on the 10-cm circle, and downloadable PDF's, see mathed.page/circle
15. This applet reveals one such connection: mathed.page/circle/sine
16. See mathed.page/alg-2/complex for games and worksheets that support this approach.
17. mathed.page/parabolas/difference
18. See a few on Henri's site (mathed.page/constructions/pythagore) and many in Steve Phelps's extraordinary collection (geogebra.org/m/jFFERBdd).

# 6
# Manipulatives

Manipulatives are physical materials, intended to be used in hands-on activities to explore or illustrate mathematical concepts.

## Models

Some manipulatives are used to model a certain domain in school mathematics. For example:

- *Cuisenaire rods* (Figure 6.1) represent whole numbers from 1 to 10. One possible use is to place them end to end to represent addition of small numbers.

**Figure 6.1** Cuisenaire rods.

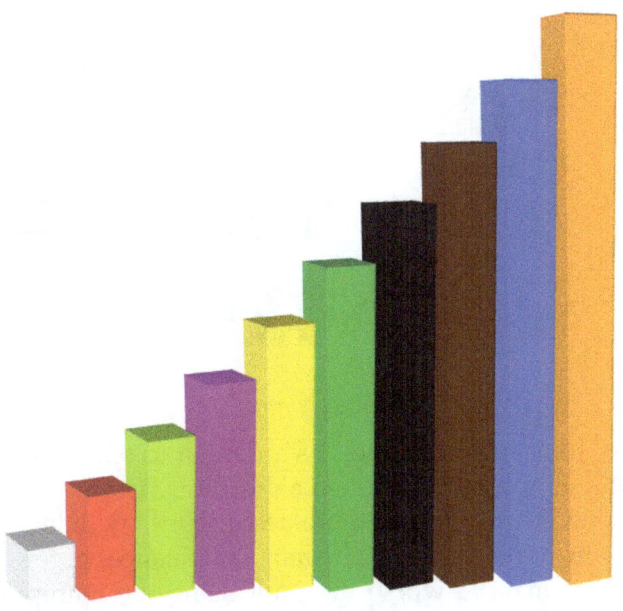

- *Base ten blocks*[1] (Figure 6.2) represent whole numbers in our base ten system: ones, tens, hundreds, and thousands. They can be used to illustrate such concepts as "carrying" and "borrowing" in calculations. They can also represent decimals, for example by decreeing that the 100 block represents 1, and therefore the 1 block represents one hundredth, or 0.01.

**Figure 6.2** Base ten blocks.

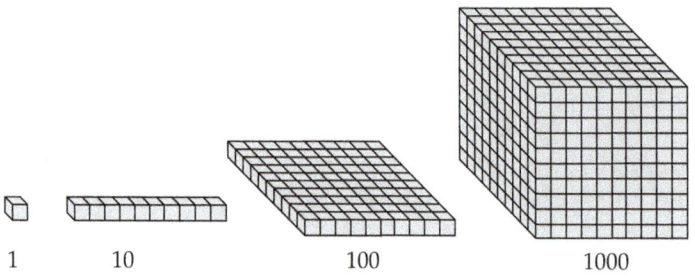

- *The Lab Gear* (Figure 6.3) is an extension of base ten blocks to algebra, with additional blocks representing $x$, $y$, $x^2$, $x^3$, $x^2y$, and so on. (There are other versions of algebra manipulatives, each with their own approach to modeling simple algebraic manipulations.)

**Figure 6.3** The Lab Gear.

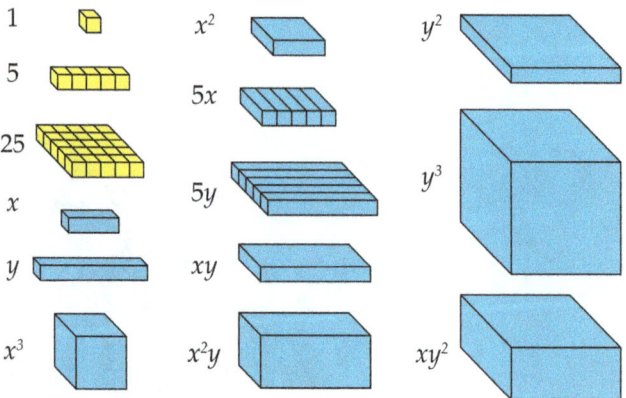

It is a mistake to use these models as an arena to teach procedures, with students imitating teacher-explained algorithms. That would only lead to students having to memorize more techniques, with only a marginal improvement in understanding, if that. On balance, given the time this would take, that may be worse than not using the manipulatives at all.

100 ◆ Classroom Practice

Instead, teachers should use the models as an arena for exploration, problem solving, and discussion.

Here are some examples:

- *Cuisenaire trains.* Here is an activity for kindergarten or first grade. $3 = 1 + 1 + 1$ can be shown with Cuisenaire rods (Figure 6.4). Is there another way to make a "3-train"? Yes! In fact, there are three other ways, shown in Figure 6.5.

**Figure 6.4** $1 + 1 + 1$.

**Figure 6.5** Three other ways.

How many ways are there to make a 4-train? A 5-train? Of course, many variations on this are possible. For example, what are all the ways to make a two-rod 10-train? This is much easier, but getting to know pairs of numbers that add up to 10 is important for mental calculation, and should be seen in multiple contexts.
- *Base ten Explanations.* Here is an activity for second or third grade. "I found that $17 + 18 = 35$. I think I made a mistake, because in 17 and 18, there were two tens, but in my answer, there are three tens. Use base ten blocks to explain my mistake, or to show me I was right".
- *Lab Gear Rectangles.* Here is a puzzle: make a rectangle, using two $x^2$ blocks, five $x$ blocks, and two ones. Write *length × width = area* for your rectangle. Figure 6.6 shows the resulting arrangement, and the corresponding symbolic representation in two formats.

**Figure 6.6** Make a rectangle.

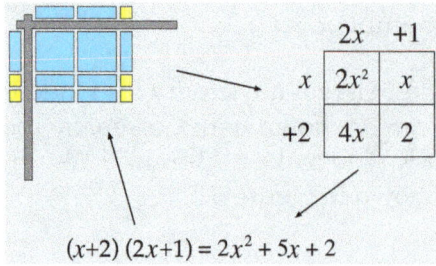

This activity is a way to take advantage of the rectangle area model of multiplication. Doing many "make a rectangle" problems of this type, and discussing them, helps lay a foundation for understanding many algebraic manipulation topics: the distributive rule, factoring, simple polynomial division, simplifying some algebraic fractions, completing the square… It is possible for all students to get involved immediately, since they know what a rectangle is. (Starting with a teacher explanation of the distributive rule or factoring would be sure to leave many students behind.) Of course, while making the rectangles provides the foundation, it is not sufficient in and of itself. Teacher-led discussions of the resulting figures are required in order for students to make sense of this.

Another, more immediate, benefit is that working with this model helps students recognize the difference between $x + 2$, $2x$, and $x^2$ as their representations are quite different from each other. The fact that so many middle school students confuse those expressions is a symptom of how the strictly symbolic approach fails to connect.

An extension to three dimensions allows for the representation of third-degree expressions. For example, Figure 6.7 shows the Lab Gear representation of $(x+y)^3$.

**Figure 6.7** Cube of a binomial.

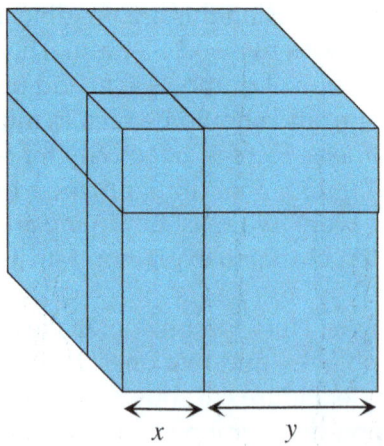

While the Lab Gear manipulatives were not originally designed for this, they also provide an engaging environment for perimeter problems, which lead to interesting basic manipulations. For example, students can be asked for the perimeter of the shape in Figure 6.8, which is made up of a 1 block and an $x^2$ block.

**Figure 6.8** A perimeter problem.

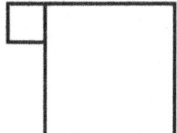

One student may answer $3 + 3x + (x - 1)$. Another: $x-1+x+x+x+1+1+1$. The best way to see if the two students agree is to simplify the expressions. And yes, both simplify to $4x + 2$, so they do agree. While gratuitous "simplify" problems are not popular among students, when they are debating the results of their perimeter problems, they care about the outcome and vigorously engage in debates about them.

An interesting follow-up puzzle is: find other Lab Gear figures with that same $4x + 2$ perimeter. Searching for those can lead to interesting discoveries and discussions.[3]

The power of using these manipulatives is that they are, to paraphrase Papert, *objects-to-discuss* and *objects-to-write-about*, as well as *objects-to-think-with*. It is especially in the ensuing conversation among students, with guidance from the teacher, that connections are made and underlying structures are revealed.

## Geometric Puzzles

Some manipulatives originate in recreational mathematics (mathematics carried out for entertainment!)[4]

### Tangrams

The original *tangram* puzzle consists of seven pieces. The goal is to create interesting representational images, such as the one shown in Figure 6.9.

**Figure 6.9** A seated tangram person.

The classic mathematical challenge is to arrange them to make a square, but it does not make a good puzzle for the classroom: it is too difficult for many students to solve in one sitting, and instantaneous for students who are familiar with it. Instead, students can be asked to create triangles, squares, and parallelograms using 1, 2, 3, … pieces. The 7-piece square is now subsumed in a broader search, one that can be engaged in by all students. Pretty soon, some students will make rectangles, trapezoids (isosceles and right), pentagons, and maybe others. This is an opportunity to introduce these terms in a natural and engaging context.

The lesson can be done with students at almost any grade level. Two extensions for high school students: Why is a six-piece tangram square impossible? For what values of $n$ is a tangram $n$-gon possible?[5]

**Pentominoes**

*Pentominoes* are the figures obtained by joining five squares edge to edge. There are twelve of them, as shown in Figure 6.10.

**Figure 6.10** The pentominoes.

Their four-square relatives (known as tetrominoes) are familiar to all who have played Tetris. (Although, unlike physical tetrominoes, Tetris pieces cannot be flipped over.) Pentomino puzzles have a long history. They are the source of many puzzles, some of them fiendishly difficult. Henry Dudeney included one in his 1919 book, *The Canterbury Puzzles*.[6] They were explored in depth and named by Solomon Golomb in his book *Polyominoes*.[7] Polyominoes were popularized widely by Martin Gardner, in several of his "Mathematical Games" columns in *Scientific American* from 1957 to 1979.

One puzzle we like is "pentomino blowups": create a scaled version of each pentomino, with double or triple the dimensions of the original, and cover it with pentominoes. An example is shown in Figure 6.11.

**Figure 6.11** Scaling the "F" pentomino.

What is interesting is that if the dimensions are doubled, the solution requires four pentominoes, and if they are tripled, it requires nine pentominoes. The general rule this illustrates is that if the dimensions are multiplied by $k$, the area is multiplied by $k^2$. (This is because a unit square in the original shape is blown up to a $k$-by-$k$ square in the scaled figure.) This rule is difficult for students to remember, even if they understand it, and having their solutions to the scaling problem displayed on the bulletin board helps to keep that concept alive.[8]

### Supertangrams

The idea can be taken one step further with *supertangrams*, which are shapes made of four isosceles right triangles, a sort of conceptual melding of pentominoes and tangrams.[9]

**Figure 6.12** The supertangrams.

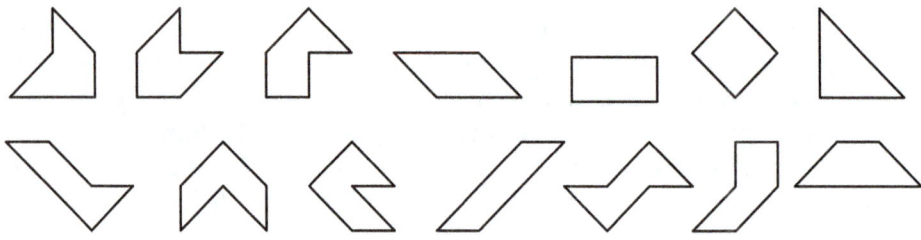

Figure 6.12 shows the fourteen supertangrams. Figure 6.13 shows a supertangram scaled to different sizes, which leads to interesting questions about the ratio of similarity in each case.

**Figure 6.13** Three sizes.

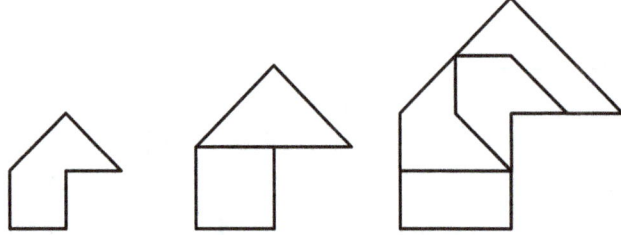

## Pattern Blocks

While puzzles are intrinsic to tangrams and the like, other unassuming manipulatives suggest their own puzzles. *Pattern blocks* (shown in Figure 6.14) can be used to explore questions about fractions, angles, area, perimeter, tiling, symmetry, and more.

**Figure 6.14** Pattern blocks.

Pattern blocks were invented in the 1960s by Ed Prenowitz of the Education Development Center. Thus, they are older than one of the authors of this book. They have their own Wikipedia entry.[10]

One example of the versatility of pattern blocks is "Angles Around a Point", an activity in which students are to surround a point with pattern blocks in as many ways as possible with each block touching the point at a vertex. Figure 6.15 shows a case where the same angle is used repeatedly all the way around, which allows the students to find the angle's measure.

**Figure 6.15** 60 degrees.

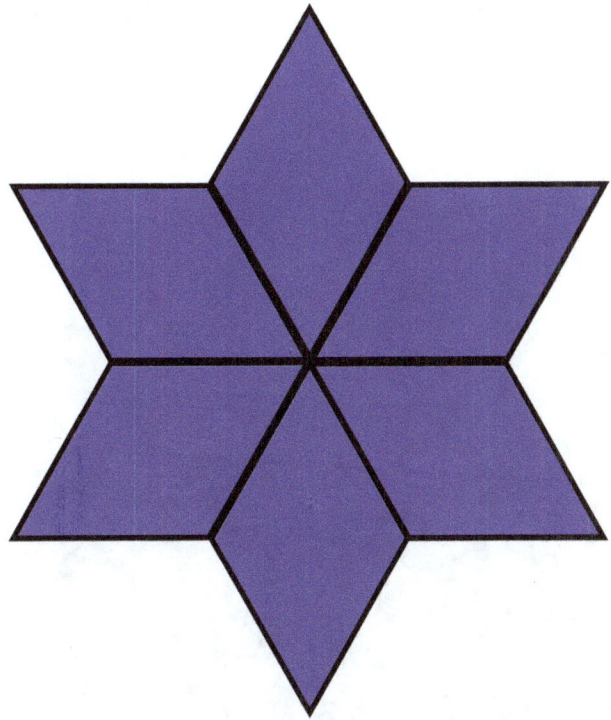

Another pattern block activity is the tiling of a regular dodecagon with pattern blocks, in search of all possible symmetries: 1-, 2-, 3-, 4-, 6-, and 12-fold rotations, with or without mirror reflections. Figure 6.16 shows two examples.

**Figure 6.16** Symmetric dodecagons.

## Geometry Labs Template

**Figure 6.17** Geometry Labs template.

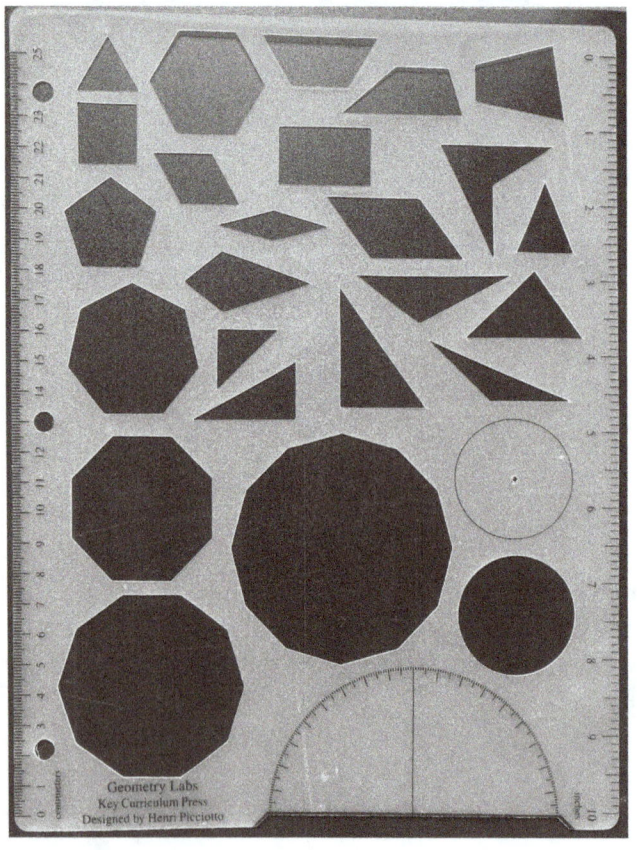

Or take a plastic *template* which allows students to draw a given geometric figure repeatedly (Figure 6.17). It is an ideal tool for the exploration of tessellations (tilings.) With the help of this template, students can discover the rather surprising fact that in fact any triangle and any quadrilateral can tile the plane (as in Figure 6.18). They can take this further by trying to find all the Archimedean tilings: the ones where all the tiles are regular polygons, and all vertices are identical – Figure 6.19 shows two examples.

**Figure 6.18** Tiling the plane.

**Figure 6.19** Two Archimedean tilings from Wikipedia.

Lest you think these designs are merely decorative, there is plenty of math to be learned or verified there. For example, the triangle tiling above can be used to show that the sum of the angles in a triangle is 180°, as well as the properties of vertical angles, and of the angles formed when a line intersects parallel lines. Tilings of the plane also suggest some basic theorems about isometries (rigid motions).[11]

A geometry template can also be used to explore the concept of a *rep-tile*, which combines tiling and scaling: are there tiles that will tessellate a scaled version of themselves? A fragment of the above triangle tiling shows that a triangle can tile a scaled version of itself with four copies of the original. Are there triangles whose scaled replica can be tiled with 2, 3, or 5 copies of themselves? Surprisingly, the answer is yes. The Pythagorean theorem can help high schoolers find them.[12]

## Even More Manipulatives

### Geoboards

Another concept from Seymour Papert is that of *microworld*. A microworld is a confined and constrained environment which is easier for students to explore than the whole world it is part of. He coined the word in reference to the turtle graphics in the Logo programming language, which we discuss in Chapter 7, but it also applies to some manipulatives, such as the *geoboard*.

The geoboard is a board featuring an array of pegs. Students can create figures on it with the help of rubber bands. Since vertices only appear on lattice points, this is a much more restricted environment than, say, a piece of paper, as the latter allows vertices to be placed anywhere.[13]

Here is an example of a geoboard activity: On an 11 by 11 geoboard, find as many different-sized squares as possible. For each one, find its area.

Students initially find ten such squares: 1 by 1, 2 by 2, and so on, with areas 1, 4, 9, ... such as the blue ones in Figure 6.20. When the teacher insists there are many more possibilities, students find "45°" squares such as the red ones. Eventually, they realize that there are quite a few more solutions, such as the remaining "tilted" squares.

**Figure 6.20** Geoboard squares.

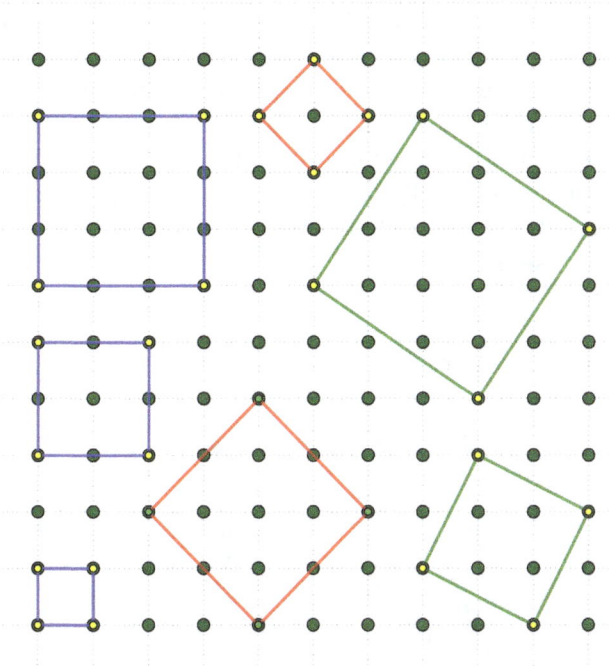

A common way students find the area of these geoboard squares is by subtraction: they encase the square in a bigger, horizontal-vertical square, and subtract the surrounding triangles from its area as shown in Figure 6.21.

**Figure 6.21** A square in a square.

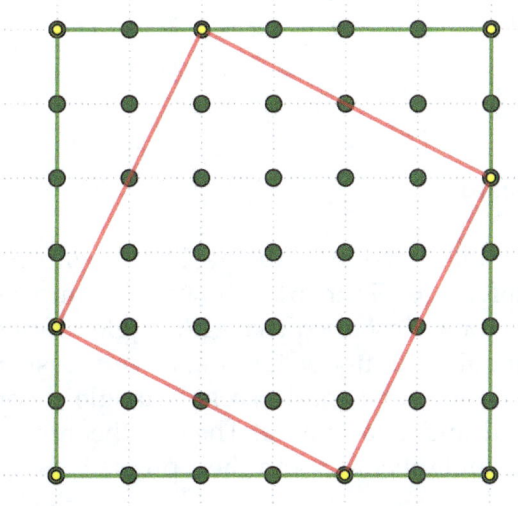

Manipulatives ◆ 111

After the students have done this a number of times, they are ready for a generalization of the calculation. As it turns out, the generalized calculation is one classic proof of the Pythagorean theorem.

The same microworld provides a great environment to learn about square roots. If students understand the geometry of square roots, they can get some insight into the operation. Consider the squares in Figure 6.22. The larger one has area 20, therefore the smaller one has area 5. but if you look at their respective sides, you see that $\sqrt{20} = 2\sqrt{5}$. This can of course be arrived at computationally, but this visual confirmation is a great conversation starter. It helps students see that the rules about simplifying radicals do make sense.[14]

**Figure 6.22** Four squares in a square.

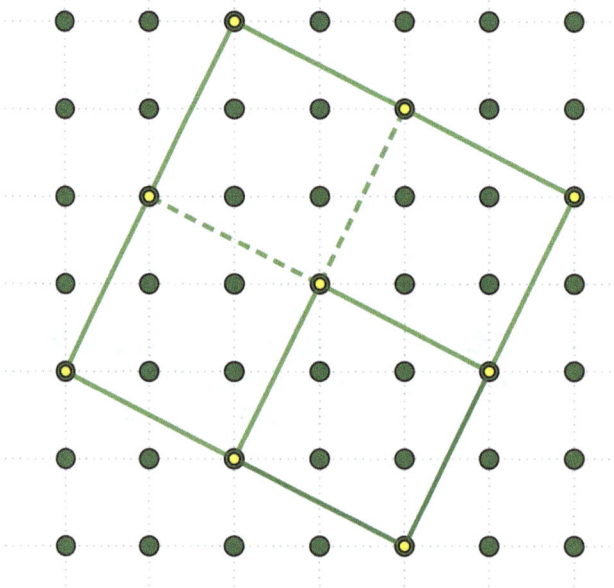

### The Circle Geoboard

There is also a circular version of the geoboard, with pegs at the center and on the circumference. Figure 6.23 shows an example of the challenge: "Find the measure of an inscribed triangle's angles, if one side is a diameter". This only requires an understanding of very basic facts: the sum of the angles in a triangle and the isosceles triangle theorem. It prepares students for a solid understanding of Thales's theorem, and more generally the inscribed angle theorem and their proofs.[15]

**Figure 6.23** A triangle in a circle.

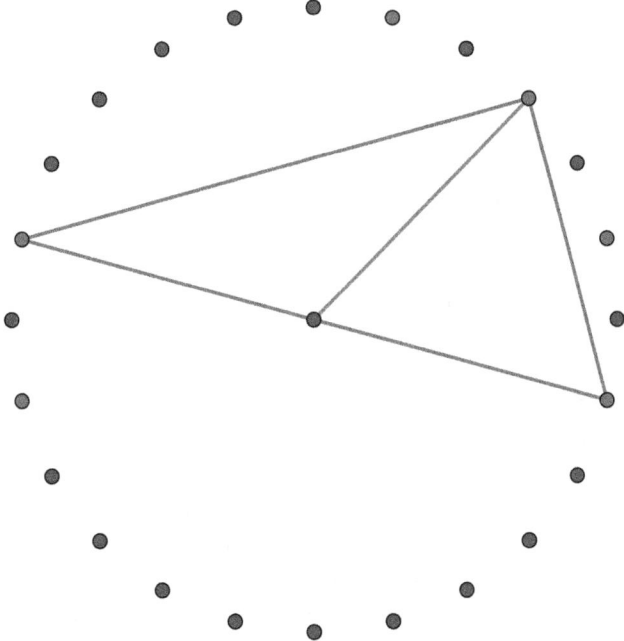

**Everyday Materials**

Of course, we have not listed every possible manipulative. For example, interlocking cubes are widely used. For more ideas, spend some time in math education publishers' catalogs! The manipulatives we did discuss were developed especially for math education. But budgetary constraints and teacher creativity have led to the use of everyday materials in the math classroom:

- Beans can be used for counting. Gluing ten beans on a popsicle stick, and combining ten such sticks into a "raft" of 100 beans is a way for little children to make their own base ten blocks.
- String can be used to measure around jars, cups, cans, etc. to find an approximation for π.
- Uncooked spaghetti can be used to fit a line to data points on a scatter plot, or broken into pieces to make triangles.
- Toothpicks can be used to create designs for visual patterns such as the one in Figure 6.24: how many toothpicks will there be in the tenth arrangement? In the 100th? In the $n$th? Such patterns can be used (in different ways) from elementary school to precalculus.[16]

**Figure 6.24** A visual pattern.

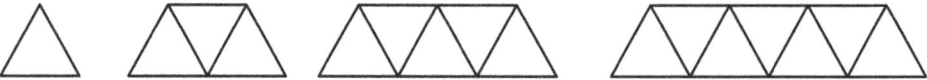

## From the Concrete to the Abstract

Mathematics is all about abstraction and logic. In fact, we consider abstraction to be a key part of the "habits of mind" meta-curriculum. However, it would be a mistake to conclude that math instruction should limit itself to abstraction and logic. Quite the opposite: it is the responsibility of educators to start in the concrete, and build from there. We do not claim that manipulatives are required for all topics, or in fact for any topic. Our claim is merely that they can provide a useful context for problem solving, for reflection, for interaction, and for discussion. It is the teacher's responsibility to help students develop the concepts from there. The manipulatives are vehicles, abstraction is the destination.

This leads us to address the "crutch" argument. If students can use algebra manipulatives for factoring, will they be hobbled when the blocks are not available to them? This is a legitimate concern, one which we address explicitly with our students, and in fact, one that is sometimes brought up by the students themselves. (Henri remembers a 9th grader passionately saying: "I want to be free of the blocks!") It is worth having a plan to gradually wean students from reliance on the manipulatives.

For example, factoring polynomials is made very accessible with the use of manipulatives, but students who are able to make a rectangle or a 3-D box with blocks do not necessarily derive any conclusions about how to do this mentally or on paper. To help that transition, the teacher should lead discussions of those exercises, focusing on the patterns therein. Then, students can be asked to factor polynomials by drawing a sketch, without access to physical blocks. Finally, the teacher can point to the connection between the geometric model and the so-called "box" representation, as in Figure 6.6. Each step in this process helps the students move from the concrete to the abstract.[17]

Another question that many teachers bring up is "Some of my students don't like manipulatives because they feel they don't need them, as they already know the material the activities are supposed to teach". Sometimes those students resist because they have been praised for their mastery of memorized symbol manipulation techniques, and they are intimidated by the visual aspect of working with manipulatives. That is often true, and such students should be encouraged and supported in developing their visual sense. But there is another issue here: if you use the manipulatives *only* as a way to provide a hands-on model of the mathematics, some students will pick this up quickly, and will feel that doing a lot of work in this environment is tedious.

This is why it is so important to include a lot of challenging/interesting/difficult work with the manipulatives, with a puzzle-maker's aesthetic (as discussed in Chapter 2). For example, with the geoboard: how many points can you put on an $n$-by-$n$ geoboard so no three are on a line? How many figures with a given area can you find? Can you represent $\sqrt{3}$ on a geoboard? And so on. And likewise with other manipulatives. Such activities are interesting to a wide range of students and, frankly, to many teachers.[18] This makes them an important ingredient in a tool-rich pedagogy. They help build your strongest students' interest and set up an atmosphere of exploration and collaboration where all can thrive.

## Discussion Questions

1. Which of the manipulatives discussed in this chapter have you used? How did it go? Which among the manipulatives you have not used are you interested in learning more about?
2. Pick an example of a manipulative and discuss it as a way to reach a particular group of students vs. as a way to reinforce a particular kind of understanding among all students. Does it work for both?
3. Manipulatives offer a physical, hands-on arena for doing math, and at the same time a visual and geometric representation of concepts. Compare and contrast how these two aspects contribute to student learning.
4. What are the arguments for and against the use of manipulatives? Think about time issues, other practical considerations, and different populations of students.

## Notes

1. Cuisenaire rods and base ten blocks are available from many educational publishers.
2. As it turns out, there is a high school use for this problem: if you look at how many ways there are to make a train, organized by how many blocks it took, you get the coefficients in the binomial expansion. In Figures 6.4 and 6.5, 1, 2, 1, so $(a+b)^2 = a^2 + 2ab + b^2$, and so on for longer trains.
3. For a lot more information on the Lab Gear, including a comparison with other algebra manipulatives, see mathed.page/manipulatives/lab-gear.html
4. Henri shares many geometric puzzles for classroom use on his website: mathed.page/puzzles/puzzles.html

5. Henri discussed the first question in *Geometry Labs* (mathed.page/geometry-labs) and the second in a blog post (blog.mathed.page/2023/04/26/convex-tangram-figures).
6. H. Dudeney. *The Canterbury Puzzles*. Dover, New York, 1958, 1986. Originally published as *The Canterbury puzzles and other curious problems*. London. W. Heinemann, 1907.
7. S. Golomb. *Polyominoes: Puzzles, Patterns, Problems and Packings*. Princeton University Press, Princeton, 1994. Originally published in 1965 by Scribner.
8. Find hundreds of pentomino puzzles and an online pentomino applet at mathed.page/puzzles/pentominoes.html
9. mathed.page/puzzles/supertangrams.html
10. Pattern blocks are available from many educational publishers, and several websites offer an online version. For their use in secondary school, see mathed.page/manipulatives/pattern-blocks
11. For tiling activities across the grades, see mathed.page/tiling.html
12. See mathed.page/geometry-labs/rep-tiles.pdf
13. For much more about geoboards, including an online geoboard applet, see mathed.page/geoboard
14. This is from one of the *Geometry Labs* on Henri's website (mathed.page/geometry-labs).
15. Henri shares an online circle geoboard at mathed.page/geoboard/circle-geoboard
16. See VisualPatterns.org for hundreds of examples.
17. This transition is illustrated in a video on Henri's website: mathed.page/manipulatives/slides/lg-1b-rectangles
18. See blog.mathed.page/2017/03/05/geoboard-problems-for-teachers/

# 7

# Computation Engines

How should technology in the world affect what we should do in the classroom? This has been a question for several decades now, but since pedagogy does not change as fast as technology, it may still be too early to give a definitive answer. There are now apps that allow a student to aim their phone at an equation, say, push a button, and get a step-by-step solution. The belief that the existence of such software should have no impact on math education is absurd on its face.

## A Brief History

Technology influences both the content and the methods of math education, but the impact is slow and gradual, not sudden and dramatic. This is in part because it takes time for technology to reach the classroom. It also takes time to sort out the effect of the new technology within the classroom, devise uses that result in educational gains rather than detracting from students' experiences, communicate experiences with the new technologies across schools, and refine lessons through trial and error.

Here are some high school topics which have vanished from the curriculum because of the impact of technology.

- How to use a slide rule. Using a slide rule effectively was connected with some math content (understanding significant digits, logarithms, and no doubt other topics). The advent of the scientific calculator relegated the slide rule to the closet or the trash can.
- An algorithm to calculate square roots with as many digits as you need. This was a long-division-like process carried out on paper. Understanding the algorithm was not a primary goal of instruction – the goal was to get the answers efficiently. Obviously,

the calculator was even more efficient, so the algorithm is no longer taught.
- Log and trig tables. One side benefit of using those was developing some skills in interpolating between values. But in the end, this was another tool that was displaced by the calculator.

In all three cases, we replaced one tool with another, and this was not particularly contentious. Yet, as technology continued to advance, some curricular implications of those developments have been quite controversial. We will discuss three specific instances, in chronological order.

**Basic Arithmetic**

We are told that schooling is intended to prepare you for the job market. And also that it is important to know how to use the traditional algorithms for calculations. Yet, there is not a single line of work that requires facility or even basic competence in paper-pencil arithmetic. This is true in retail, health care, manufacturing, finance, engineering, etc. Everywhere in society (outside of schools) arithmetic calculation is done electronically, and students know it.

We are told that using the algorithms helps students develop number sense. This is preposterous: the whole point of these algorithms is to facilitate fast and accurate computation. It has nothing to do with understanding numbers. Quite the opposite: thinking about numbers is a distraction from carrying out the step-by-step instructions, which need to be 100% automatic. This is why it is crucial, in that view, to automate basic addition and multiplication facts, irrespective of the math that is embedded in them.

And yes, there is much interesting math embedded in the addition and multiplication tables, long division, and so on. It is useful and interesting to learn that math, and it is best accomplished through number talks, base ten blocks, Cuisenaire rods, one-on-one conversations with students, and so on. Analyzing algorithms, including the traditional ones, also offers opportunities for interesting mathematics. But aiming for speed and accuracy in paper-pencil arithmetic just does not make sense anymore. It takes a lot of time spent in mindless practice, and it is so boring and frustrating that it turns many students off to mathematics.

We can get at many important ideas by moving mental arithmetic to center stage. And yes, some facility with basic addition and multiplication facts is helpful. It is best acquired in conjunction with thinking about the numbers. Sharing, discussing, and practicing various mental arithmetic and estimation strategies is a powerful way to develop number sense, and at the same time, it is vastly more useful in daily life than paper-pencil algorithms – in and out of the classroom. We gave an example of mental calculation in the Chapter 3 discussion of number talks. That routine can be the foundation of a discussion of the distributive property: to figure out 3 × 19, we can use the fact that

$$3 \times (20 - 1) = (3 \times 20) - (3 \times 1)$$

Likewise, estimation skills can be developed by number talks about the approximate value of calculations such as 31 × 19.

Meanwhile, calculator use in word problems will not harm students' understanding, and can in fact be combined with the use of mental arithmetic and estimation.

**Equation Solving**

Many math educators consider the solving of linear equations to be the heart of algebra. The topic is often introduced early, atomized into many special cases (one-step, two-step, etc.), and the subject of endless mind-numbing practice. For many students, this carries little meaning, other than "I'm supposed to get $x$ by itself". This is counterproductive. Algebra is about many things: structure, including especially the distributive law; modeling real-world phenomena; functions; exponents; and so on. Equation solving is indeed important, but it is much better taught in concert with all these other dimensions of the subject. And note that of all conceivable equations, only linear and quadratic equations can be solved with paper and pencil. If exponential, trigonometric, or logarithmic expressions are involved, Algebra 1 techniques are useless.

On the other hand, any equation in one unknown can be solved graphically. This was already true once the graphing calculator entered the scene, and is only more so in the age of Desmos[1] and GeoGebra.[2] We suggest that for equation solving, graphical solutions be given equal status to traditional algebraic methods.

Probably the biggest loss resulting from such a policy would be that students would get less practice in symbol manipulation. Here are two ways to make up for that. First, we suggest equation talks: solving linear equations mentally, and discussing what strategies make that possible. That would help center meaning, it would help the teacher get a sense of student understanding, and it would reinforce the idea that there are many ways to proceed.

Second, we suggest more work on equations with more than one variable, and with literal equations. For example, if $e = mc^2$, solve for $m$. Those are quite useful in manipulating formulas in science classes, but they are notoriously difficult for students. Part of that is because so little time is spent on them. On the way to that, perhaps lessons could be developed where students solve for $x$ in equation batches such as this one:

$$2x + 3 = 7$$
$$2x + b = 7$$
$$mx + 3 = 7$$
$$2x + 3 = y$$

The gains of promoting graphical/electronic equation solving would be substantial. It would increase access, as more students would readily be able to use those techniques. It would make room for more applied and modeling problems, which would help increase motivation. It would require a better understanding of graphing and functions, which are crucial in subsequent courses. And it would not preclude the use of well-chosen paper-pencil problems in reasonable amounts, as well as lessons that focus on making the connection between graphical and symbolic methods.[3]

In general, students would need to know which approach is most appropriate for a given problem. For example, if a problem requires an exact solution, such as $\sqrt{2}$ or $\frac{1+\sqrt{5}}{2}$, a graphical solution would not be the right choice.

**Fundamental Concepts**

The fact that too much time in Algebra 2 was spent on disembodied manipulations has led some in our profession to call for the elimination of the course, or its exclusion from graduation requirements. Going down that path will end up increasing educational inequalities, as the children of the well-off will still have access to Algebra 2, which is prerequisite to so many topics in math, science, engineering and statistics. In our view, we should continue the work we have been doing to bring math to all students. Going all in on technology and mental math is a crucial part of that mission. It can help with understanding, motivation, and equity. Unfortunately, this requires costly equipment, and there is a huge digital divide between rich and poor schools and districts.

The concepts underlying the topics discussed in this section are still fundamental. Technology and mental math can help preserve the conceptual content, transmit necessary skills, and cut the disembodied manipulations. Long division may die, but division is not going anywhere. Equations can be solved by machines, but they remain essential in STEM fields. Understanding the distributive law is needed if students are to use formulas and symbolic representations in any quantitative discipline. However, these concepts should not be confused with the way they were addressed at different stages in the history of technology.

The task facing us is to figure out how to teach the concepts, while simultaneously deemphasizing paper-pencil manipulation. One of us remembers using a slide rule in science classes. It was fun and did require adequate number sense. Still, we do not think the slide rule should have remained in the curriculum. In our own lifetime, calculators displaced paper-pencil calculation of square roots, scientific calculators replaced logarithm and trigonometry tables, graphing calculators superseded tedious graphing by hand, and we are now in the fourth phase of this revolution: automated algebra. Of course, this does not mean that we know what to do about the new state of affairs. Many questions remain, but they will not be answered by trying to continue business as usual. We grapple with some of these questions in the remainder of this chapter.

**Graphing Technology**

A concept-first approach to equation solving begins with the question of what an equation "looks like", whether we believe it should have a solution or solutions, what we can argue about these, and finally, whether there are cases in which we can precisely compute the solution(s). For example, should the equation $x - 1 = \sqrt[3]{x}$ have zero, one, or more than one positive solution? Graphing the left and right sides (Figure 7.1), it would appear that there is precisely one solution.

**Figure 7.1** Graphing two functions.

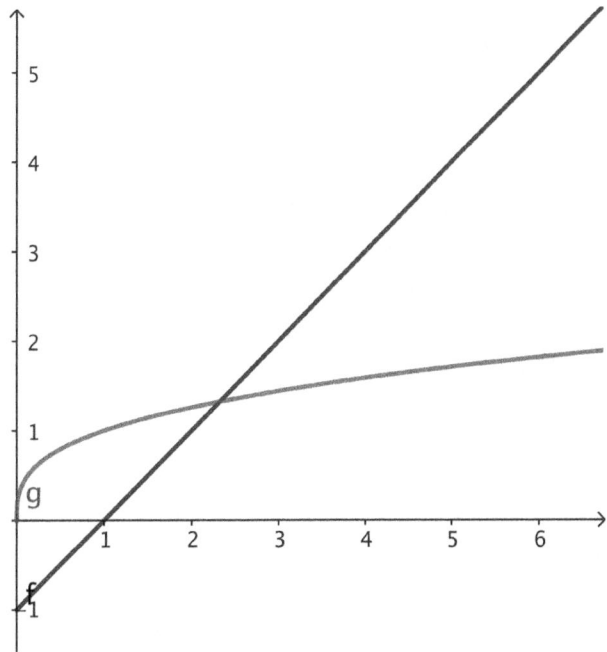

The extent to which we can argue this is correct depends on the level and sophistication of the students. It is hard to imagine, though, that any student who has heard of a cube root and seen these graphs will remain unconvinced. It may surprise some of them that the teacher can't write down a formula for this solution any better than the students can.

The existence of the calculator requires us to shift our emphasis to mental calculation and estimation. If complicated calculations are needed to solve a particular problem, we do not object to calculator use. A similar debate about calculator use has erupted in reference to graphing technology in high school. We have similar views about it. Students should be allowed to use it as they see fit, except when the teacher rules it out for a particular lesson or assessment.

Routine use of graphing technology, whether on a calculator, a phone, a tablet, or a computer makes it possible for students to quickly switch between representations: graphical, numerical, and symbolic. Connecting the representations can give students a better understanding of each one. Even at an Algebra 1 level, for example, it is easier to teach students about solving systems of linear equations if they have thought about it in those three representations: symbolic, numeric, and (Figure 7.2) graphical.

$$\begin{cases} x + y = 3 \\ 2x - y = 3 \end{cases}$$

| $x + y = 3$ | | $2x - y = 3$ | |
|---|---|---|---|
| $x$ | $y$ | $x$ | $y$ |
| 0 | 3 | 0 | −3 |
| 1 | 2 | 1 | −1 |
| 2 | 1 | 2 | 1 |
| 3 | 0 | 3 | 3 |
| 4 | −1 | 4 | 5 |

**Figure 7.2** Graphing a system of linear equations.

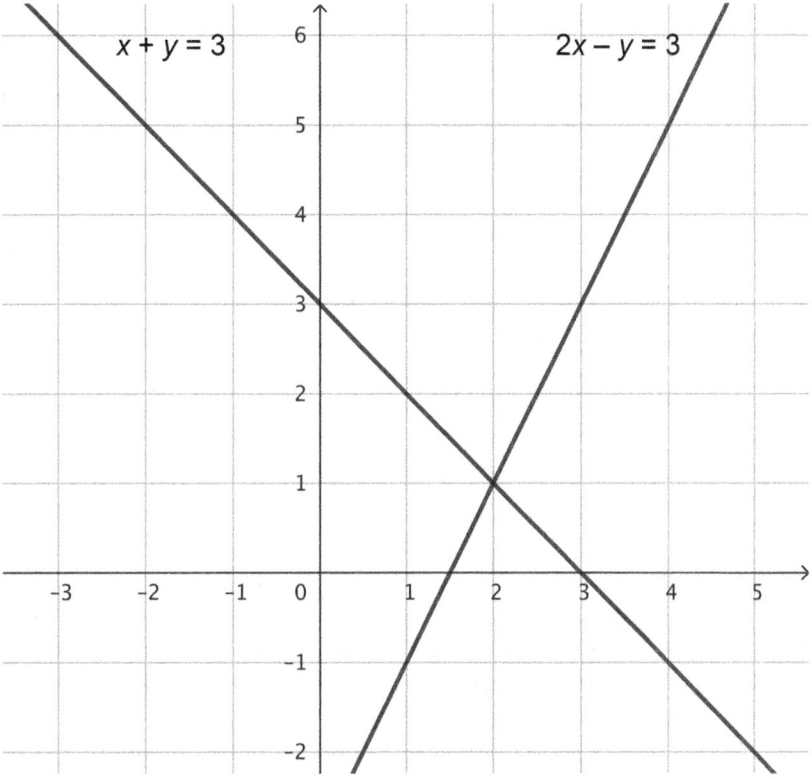

The risk is that if students can obtain a graph simply by typing into a graphing utility, they will not learn crucial underlying concepts about graphs, such as intercepts, asymptotes, or the long-term behavior of the function. However, that risk cannot be pinned on the technology: it is also a risk if they *only* learn to graph functions the old-fashioned way, by graphing individual points on paper, and connecting them with a smooth curve.

In short, these concepts must be taught. For example, one can find out a lot about the graph of $\frac{x^2+1}{x-1}$ by inspecting the expression. It is not defined for $x = 1$, and gets very large when $x$ is close to 1, as the numerator approaches 2 and the denominator gets very small. So, $x = 1$ will be a vertical asymptote, and $y$ will approach negative infinity if $x$ approaches 1 from the left, and positive infinity if it approaches 1 from the right. As for the function's behavior as $x$ gets very large (positive or negative), the ratio of the $y$ value to the $x$ value will approach 1, because the effects of the +1 in the numerator and the −1 in the denominator become negligible.

Computation Engines ◆ 123

Mathematical modeling often starts with the ratio frame, where this observation is satisfactory, but in this case, more precise information is available. Rewriting the identity as $y = x + 1 + \dfrac{2}{x-1}$, we can see that the diagonal line $y = x + 1$ is an asymptote, that the graph approaches the asymptote from above as $x \to \infty$ and below as $x \to -\infty$, and that the rate of approach is asymptotically $\dfrac{1}{|x|}$. We can also tell that the function will cross the $y$-axis at $(0, -1)$ and that it will not cross the $x$-axis, because the numerator is never 0.

Depending on the level of the students and how the math is embedded in a problem, any of these points might be the crux of the matter. The most important takeaway might be that there are many interesting things to say about this function, and that continuing to mine information gives a better picture of what's going on.

All of this can be learned, whether or not students have access to graphing technology. Nothing prevents the teacher from working on these concepts in no-tech lessons, and assessing understanding in no-tech quizzes. Or a rule can be made for a particular lesson to only use the electronic grapher for checking the correctness of answers. On the other hand, if the teacher believes those ideas are too hard for their students, then their prophecy will be fulfilled, because they will not teach this – whether technology is available or not.

But graphing technology can offer much more to the algebra class than answer-checking. For example, a discussion of these functions, before and after graphing them, can yield important insights:

$$y = \sqrt{x}$$

$$y = \sqrt{x^2}$$

$$y = \sqrt{4x^2}$$

$$y = \sqrt{\dfrac{x^2}{4}}$$

$$y = \sqrt{x^2 + 4}$$

There is a lot to see and discuss: many students are surprised that $\sqrt{x^2}$ does not always equal $x$. After understanding that, they enjoy seeing their predictions about the third and fourth graphs confirmed, but many are surprised that the final function does not look anything like the previous three.

Of course, graphing technology can be used poorly, as in this scenario:

**Teacher:** "Graph $y = x$, $y = 2x$, $y = 3x$, $y = 4x$. What do you notice?"
**Student:** "The line is getting jagged".
**Teacher:** "Graph $y = 4x+1$, $y = 4x+2$, $y = 4x+3$, $y = 4x+4$. What do you notice?"
**Student:** "The line is moving to the left".

True, and possibly the start of an interesting conversation about screen resolution, or the x-intercept, but probably not what the teacher hoped for. What students notice is constrained by what they already know and understand. In this scenario, the teacher (or the worksheet) is choosing the functions, basically doing all the work, and the student is only slightly more engaged intellectually than in a traditional "graph this, graph that" activity.

A more promising approach is to ask students to make the designs in Figure 7.3 by graphing linear functions.

**Figure 7.3** Make these designs!

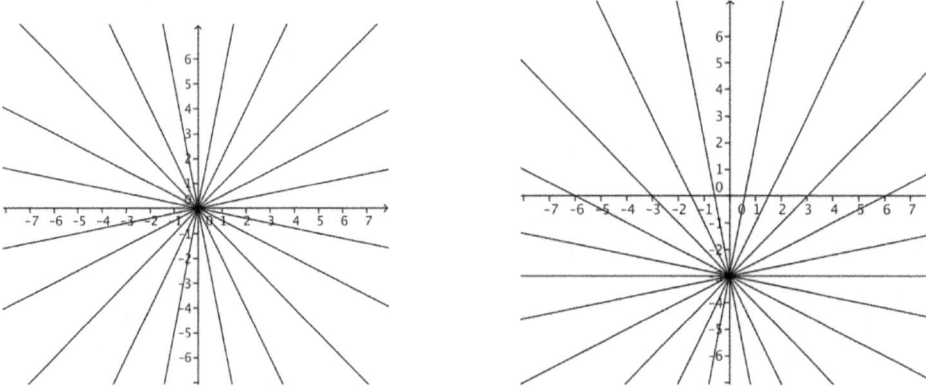

This activity can be used to practice and reinforce the slope-intercept formula for linear graphs, or to review it. Because students create the needed formulas, they are forced to think about the meaning of the parameters. To make sure they do not tune out intellectually and fall into mindless trial and error, they should be asked to write up an explanation of how each design was created: what slopes did they use? What intercepts? What patterns did they rely on?[4]

The essence of this activity is turning a traditional (and boring) assignment (graph this function) into an engaging and thought-provoking one (what is the function that yields this graph?) It can be asked at any level: make these parabolas, make these periodic graphs, make these graphs of polynomials…

Moreover, the activity can be extended: students can create their own designs, especially once they know how to use the technology to restrict the domain of the function. For example, by superposing pieces of quadratic functions, a student used Desmos to create the image in Figure 7.4.[5]

**Figure 7.4** A quadratic turtle.

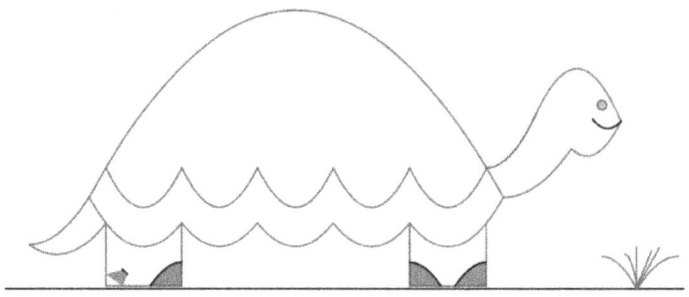

"Desmos art" projects are being assigned by more and more teachers.

**Electronic Graphing Puzzles**

A higher tech version of "Make These Designs" is a video game of sorts, in which students must get a graph to pass through certain items on the screen. In the early days of the personal computer, there was a game called "Green Globs" where the targets were randomly placed blobs. A more recent implementation is "Marbleslides" on Desmos. In that game, stars are the targets, and marbles slide down student-created graphs. (There is gravity pulling the marbles downward.) The targets are not random: they are placed in strategic locations in order to encourage students to use specific properties of certain graphs when they create their slides. Figure 7.5 is an example involving a quadratic function.

**Figure 7.5** Marbleslides.

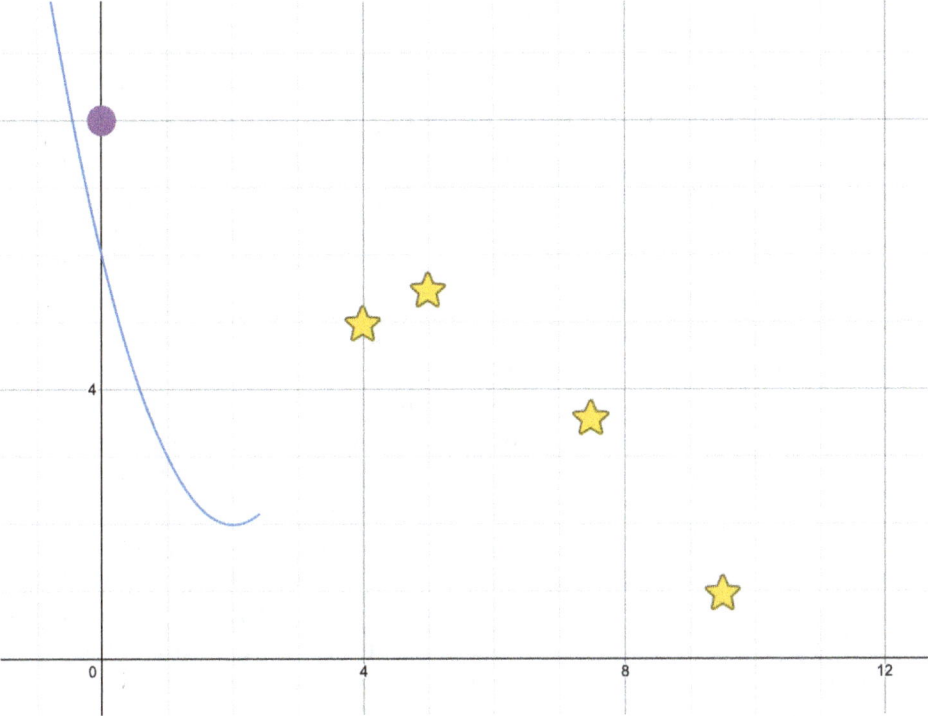

Desmos goes much further than previous electronic graphers in that it is supremely user-friendly and popular with students. It makes it possible for teachers to create multistep activities and share them on the Web. Crucially, it facilitates classroom discussion by allowing students to share their answers and hypotheses with the whole class, under pseudonyms.

## Interactive Geometry

Everyone supports the use of compasses, rulers, and protractors in the teaching of geometry. We like those, but we would like to expand the geometric construction toolbox. We are probably not the only ones who identify as too uncoordinated for these traditional tools. Henri writes:

> Back when I was a high school student, I had mixed feelings about compass and straightedge constructions. On the one hand, I liked the geometric challenge, on the other hand, I hated the physical challenge of working with an actual compass. Maybe 20 years later, I had exactly the same experience as a high school math teacher.

I loved construction, but I hated managing the compasses. Having to help the students whose papers tear or whose compasses' openings change even as they draw their arcs mirrors and multiplies the problems I faced when I was a student.

One way to reduce physical compass use is to use other tools alongside it or instead of it. Inexpensive tracing paper is available in the form of "patty paper", which fast food restaurants use to separate hamburger patties. Patty paper provides a simple way to copy figures. Perpendiculars and bisectors can be created by folding.[6] Plexiglass see-through mirrors[7] offer another way to bisect angles and segments. These tools complement the traditional ones, and are easier to work with, as they rely on and develop intuitions students already have.

But the most substantial development in this domain is the availability of interactive geometry software. Early pioneers in this field were Cabri and Geometer's Sketchpad. The current standard is GeoGebra. We have found that this technology enhances learning in a variety of ways. Direct manipulation on a screen is an excellent environment to distinguish the essential properties of a figure from the particular way it appears on a static page. It makes it easy to generate conjectures, and to find counter-examples. It is what finally might help fulfill the promise of geometry as an appropriate arena for an introduction to proof for a wide range of students. It is also a fine tool for teachers and curriculum developers to quickly create applets and microworlds to introduce powerful ideas.[8] Finally, this technology is more accessible than ever, as it is now available with free software that will run on tablets as well as computers.[9]

The essential mathematical concept underlying geometric construction is not the use of straightedge and compass. Interesting construction challenges have been developed for straightedge and the collapsing compass (a compass that does not "remember" its opening), for the compass alone, and for interactive geometry software. There are even interesting challenges involving only a (two-sided) straightedge which allows one to readily create parallel lines.

The essential concept underlying geometric construction is that of intersecting loci. As mentioned in Chapter 3, the locus of a point is the set of all possible locations of that point, given the point's properties. Typically, it is a line or a circle. If one knows two loci for a certain point, the point must lie at their intersection. In other words, given a figure, an additional point can be added to it in a mathematically rigorous way by knowing the locus of the point in two different ways. Geometric construction is the challenge of finding such points and, in some cases, using them to define additional parts of the figure.

For example, the standard straightedge-and-compass construction of the perpendicular bisector is based on this theorem: *PA = PB if and only if P is on the perpendicular bisector of AB.* (We mentioned a kinesthetic introduction to this theorem in Chapter 3.) The construction (shown in Figure 7.6) starts by using a compass to find all points at a distance AB from A: a circle centered at A, with radius AB. That is one locus. Similarly, the circle centered at B with radius AB is the locus of points at a distance AB from B. These circles intersect at two points. Each of these points is equidistant from A and B, and therefore must be on the perpendicular bisector of AB. These two points uniquely determine a line, which is the bisector.

**Figure 7.6** Perpendicular bisector.

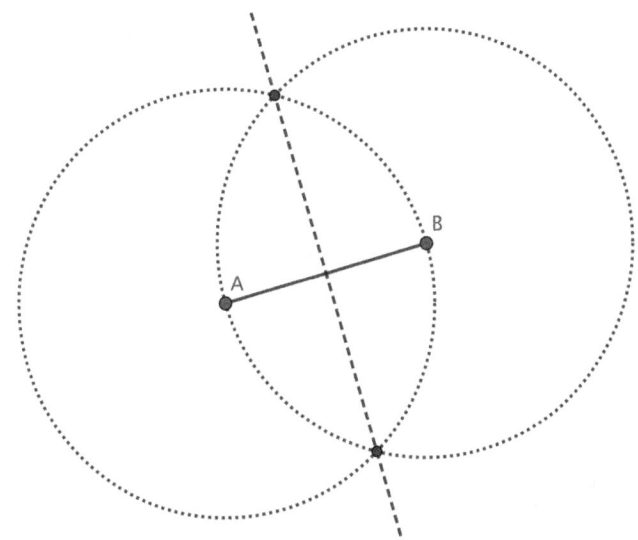

To generalize, the essential construction question is: "Given this figure and these tools, construct these additions to the figure in a mathematically rigorous way". In other words, it is a problem to solve, not a recipe to execute. In this view, the student is a thinking human being, not a programmable machine. The pedagogical question becomes: what tools are most effective if we want to use geometric construction to teach geometric concepts in part through student problem-solving?

The answer need not be limited to straightedge and compass, but whichever tools you settle on, there is an educational overhead cost: students have to get familiar with the tool before they can use it effectively. In our experience, it is well worth it. In particular, interactive geometry provides us with an opportunity to get away from the procedural approach to geometric construction ("this is how you bisect a segment; this is how you construct a perpendicular, etc."). It allows us to get back to construction as it is meant to be: mathematical challenges of increasing complexity, which require and help to develop geometric intuition and understanding.[10]

## Computer Algebra Systems

Computer algebra systems (CAS) can be used to carry out all the basic procedures of algebra, trigonometry, and calculus, including factoring polynomials, solving equations, finding the derivatives and integrals of many functions, and so on. They have become widely available on the Web, as well as in applications for computers, tablets, and phones. We saved this for last because, unlike other technological tools, the educational implications of CAS are only beginning to be explored, even though the technology has existed for decades.

GeoGebra has a CAS built into it. Figure 7.7 shows an example.

**Figure 7.7** Automated factoring.

$$\text{factor}(x^2 - 5x + 6)$$
$$\rightarrow \quad (x-3)(x-2)$$

It follows that we can spend much less time on factoring (as a skill) and put some of that time into understanding factoring, as a concept. Part of that is accomplished with the rectangle model, as discussed in Chapter 5. We're especially distressed when we hear of teachers spending an enormous amount of time teaching arcane factoring techniques and shortcuts. Some trial-and-error factoring of trinomials is a good thing, as it helps reinforce understanding of the distributive property. Beyond that, time is better spent on other things.

What about equation-solving? Photomath is a free smartphone app which can "read" exercises (even handwritten ones), solve them instantly, and display one or more paths to the answer. Here is an example. We handwrote a system of equations, and aimed the phone at it. Figure 7.8 shows the result.

**Figure 7.8** Automated equation solving.

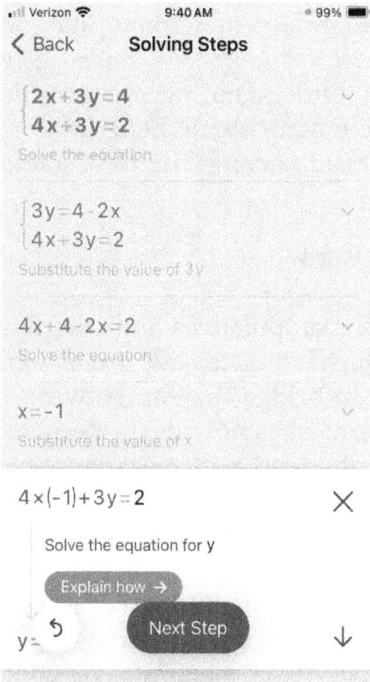

Each step can be expanded to show more details. Scrolling down gets you to the solution, and then you are shown how to check whether the answer is correct, once again with the option to expand each step. And that's not all! The app displays a graph of the two lines (and for some reason their $x$- and $y$-intercepts).

Should we explain to the students they can only use such a tool effectively if they understand the underlying math? or should we take advantage of it to assign different, more interesting problems? In our view, we should do both.

Take basic linear equations, the subject of a lot of deadly drill in middle school. Students need to know what it means to solve an equation, but they do not need to be able to solve super-complicated examples. One way to teach the basic underlying concepts is to solve a lot of equations mentally, perhaps using a number talks format (as presented in Chapter 3). "If $3x = 18$, what is $x$?" and increase the difficulty from there. ($3x + 2 = 20$, $3x + 2 = 21 + 5x$, and so on.) Another way, once the very basics have been established, is to ask questions like "create an equation whose solution is 6". This is a good way to consolidate understandings about "doing the same thing to both sides". Another advantage of such a challenge is that there are more correct answers than children in any class, so that each student can find their own solution. And of course, there are word problems, modeling questions, and assorted applications, none of which can (yet) be solved reliably by machine.

Even when it comes to algebraic manipulation, *speed and accuracy in paper-pencil computational manipulations can no longer be priorities in math education*. Teaching for understanding is really the only game in town. Trying to teach the same algorithmic material the same way as the technology keeps racing ahead becomes more obsolete every day.

## Virtual Manipulatives

Somewhere in between manipulatives and computation engines, we have *virtual manipulatives*, which simulate hands-on materials on a screen. Virtual manipulatives attempt to bridge the gap between high-tech and low-tech learning tools. That attempt has not always been successful. In some cases, the tools are just poorly designed, perhaps because the designer didn't understand how the manipulatives are actually used. (For example, pattern blocks that rotate in 5° increments, so that it takes six clicks for the smallest useful rotation.) In other cases, the design was decent but did not use the power of the computer, so that switching to an electronic version of the manipulatives led to a loss in classroom interaction, with nothing gained. Still, in a time of remote instruction, virtual manipulatives could not be compared with physical manipulatives, since those are typically not available in students' homes. The comparison was with not using manipulatives at all.[11]

Besides, virtual manipulatives help with whole-class discussions, since they can be projected. They are also handy in the preparation of worksheets (or this book!), as they can yield screenshots that can be inserted into word processor documents, and in some cases make it easy to annotate figures when using a tablet. For example, in Figure 7.9 we illustrate two activities we discussed in Chapter 6.[12]

**Figure 7.9** Using virtual geoboards.

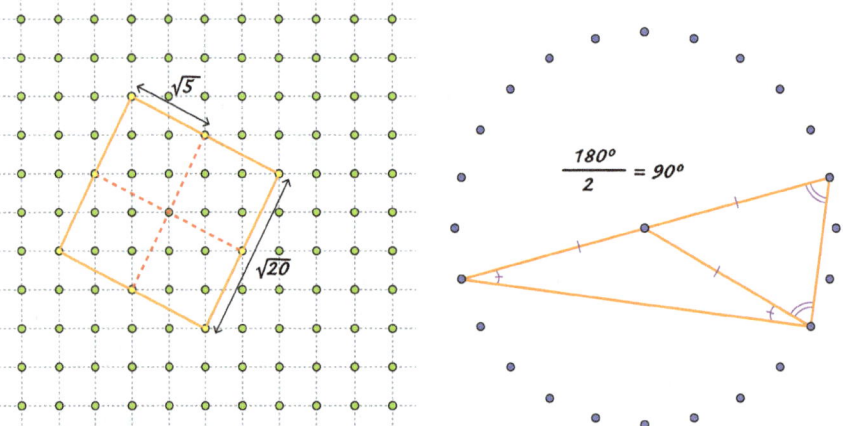

Where virtual manipulatives shine is in cases where they make possible things that cannot be done, or cannot be done as well, with physical manipulatives or computation engines. For example, one can use the power of computers to quickly create symmetric figures or tessellations with virtual pattern blocks. Figure 7.10 shows two examples that involved repeated use of GeoGebra's rotation and reflection tools.

**Figure 7.10** Pattern block symmetry.

## Tools as Curriculum

So far, we have discussed electronic tools whose main purpose is education. But in fact, computational engines have become so central to society that they have much broader relevance to business, science, and even everyday life. As a result, some of those tools themselves may be considered a legitimate topic in curriculum. Here we are thinking for example about statistical packages, graphics software, spreadsheets, and computer programming.

As long as we harness these tools for specific math curricular goals, that is not in conflict with our mission as math teachers. Because the mathematical assumptions underlying statistical software are well beyond high school math, we'd rather see this tool in the hands of science and social studies teachers, because we cannot make it understandable in the same intellectually honest way that we try to use when teaching math concepts. In contrast, key features of graphics software involve essential high school geometry content, and thus teaching those tools can profitably complement curricular work on geometric transformations.

We discuss this in more detail in the case of spreadsheets and computer programming.

**Spreadsheets**

Consider this problem:

> Paul's Forestry Products own two stands of trees. This year, there are about 4500 trees in Lean County, and 5500 in Cool County. So as to not run out of trees, their yearly harvesting policy at each location is to cut down 30% of the trees and then plant 1600 trees. What will happen in the long run?[13]

A spreadsheet is an excellent environment to explore this problem. Students are surprised to find that after many years pass, that company will end up with about 5333 trees in both counties. Would the same thing happen if they started with different numbers? (Yes!) Why? Can we generalize to find out what would happen with different parameters? This problem would be tedious and/or inaccessible without technology. A spreadsheet exploration yields interesting conjectures, and begs for a general explanation of what happens. This problem provides a basic introduction to dynamical systems.

Spreadsheets are also useful in learning about standard middle school topics, such as mean and median, and high school topics, such as arithmetic and geometric sequences and series. If a math teacher is asked to teach the use of spreadsheets, it is a win whenever the lesson can be designed also to teach a curricular topic.

**Programming**

We can argue, as others have,[14] that computer programming is a literacy, because it can enhance one's learning in other disciplines, and because it supports such uses as the creation of macros and scripts to enhance one's use of word processors or spreadsheets.

Turtle geometry is an activity that can teach programming while also enriching math education. It is engaging. It lends itself to experimentation, rapidly yielding fascinating images. At a beginning level, it is a way to explore angles and polygons. At a more advanced level, it provides an environment to explore fractal geometry. The graphics in Figure 7.11 were programmed by high school students.

**Figure 7.11** Student-programmed images.

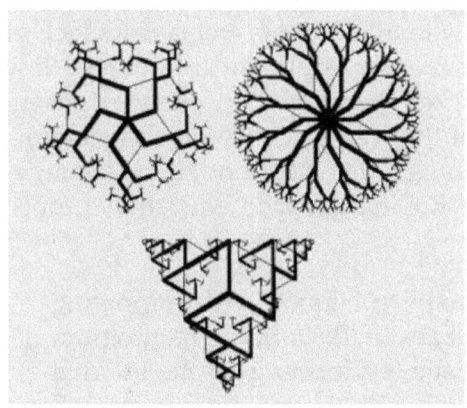
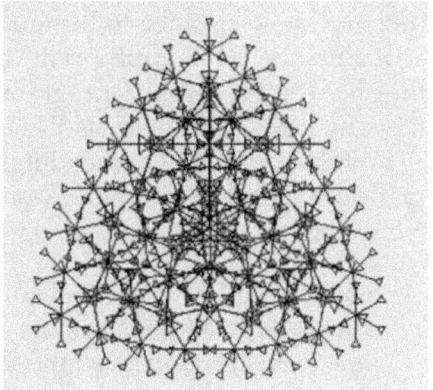

A single program generated the images on the left, which illustrate a tree structure, combined with rotational symmetry. The image on the right deserves close examination, as it includes an overall three-fold symmetry, but locally has four-, five-, and six-fold symmetric substructures. These examples show a combination of creativity and rigor rarely seen in math classes that exclude programming. To make these images, the students were not just tweaking the parameters of existing code for picture generation. They were programming from scratch in a computer language that supported turtle graphics.[15]

Our point here is that the right programming environment allows a third grader to be creative, while also offering possibilities for college-level explorations. This combination of access and challenge started with Papert's Logo. While we do not believe all math learning can happen in this format, we acknowledge that Papert's pedagogical theory is a significant contribution to math education.[16] Logo has had many descendants over the years. One contemporary example is MIT's Scratch[17] (for elementary school) and UC Berkeley's Snap![18] (for high school).

Probability yields many activities that can be done while learning to use a spreadsheet or basic coding. Usually introduced first with physical coins, dice and cards, probability experiments with those quickly become tedious. Programming makes it possible to experiment at a scale where the laws of large numbers can be seen in action.

Some mathematical ideas are algorithmic by their very nature and can be taught better with the help of programming. Mathematical induction, for example, is not unlike the sort of iteration that is fundamental to programming. Or take Riemann sums, a basic ingredient in calculus: those can be created and manipulated by way of student-written computer programs.

We don't claim that programming is necessary to teach math well, only that it can be helpful. The programmer's habits of mind are useful in a wide range of human activities: breaking down large tasks into small ones; thinking about structure hierarchically and logically; seeking clarity in presentation and precision in language; and so on. We suspect that much of the resistance to bringing this to all students stems from lack of acquaintance with appropriate curricular examples in a suitably "low threshold, high ceiling" programming environment. (In our view, using the languages professionals use is not the way to go in the classroom.)

While on the topic of coding, although this does not address mathematical learning, we mention one of Robin's favorite educational tools. He writes:

> Shamelessly low-tech, the Cardboard Instructional Aid to Computing (CardIAC) was produced by Bell Labs in 1969. Students program in a 10-instruction machine language, performing all the internal steps that a computer would perform, with the aid of a bug-shaped movable pointer, recording data into a 100-cell memory with pencil and eraser on rewritable glossy cardboard. The CardIAC is an especially good vehicle for teaching about the dual nature of data and code, which turns out to be important later in the field of logic (and also in the study of DNA).

Figure 7.12 is a CarDIAC replica built by Michael Gardi; the original is at the Drexel University Museum.

**Figure 7.12** The CardIAC "computer".

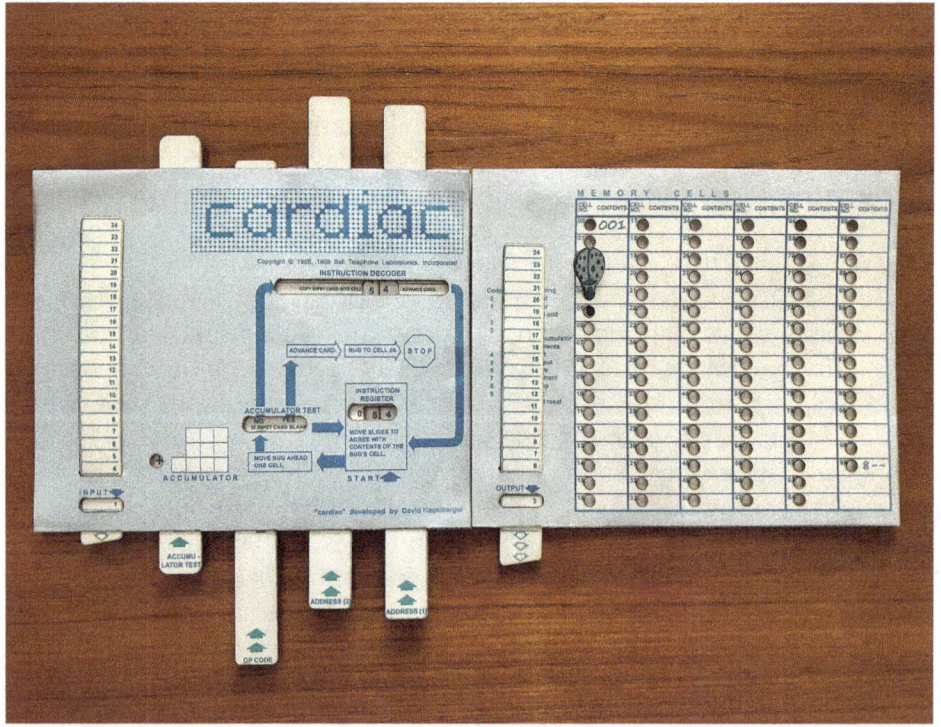

## Robot Teachers?

We cannot end a chapter on technology without discussing the idea that teachers can be replaced by machines. This takes two forms: video lectures, and supposedly "intelligent" learning programs.

Watching videos cannot replace live classroom interaction. In the next chapter, we discuss how a highly interactive approach to whole-class dynamics is much more engaging and effective than a one-way lecture. On the other hand, some use of video lectures can be productive in the so-called "flipped classroom" paradigm. The idea is to maximize student involvement with problem solving during class time when peers and the teacher are available to support student explorations. In order to do that, necessary teacher explanations, or teacher step-by-step solutions to challenging problems are relegated to at-home viewing. This setup allows students to watch such videos as many times as necessary and preserves class time for intellectual engagement. Such videos are not hard to find on the Web, but this works better if the videos are recorded by the students' own teacher, who knows where the class is at any given time.

Learning programs keep track of student progress and present them with exercises calibrated to their individual needs, as diagnosed by their performance on previous exercises. By their very nature, they tend to work best in narrow practice of basic skills. They cannot support the overall goals we set in Chapter 1, or support student engagement in the sort of problem solving we recommended in Chapter 2. Nor can they offer the rich variety of teaching modes we recommend in Chapters 3 and 4. Finally, they typically exclude work with manipulative and even technological learning tools. Thus, they may have a limited role in an overall program, but they should not be overestimated.

At least for now, live teachers remain crucial to the teaching of math. They can coordinate the use of multiple tools. If necessary, they can make in-the-moment decisions to redirect a lesson. They can perceive underlying misconceptions in a student's incorrect solution. They can read expressions on students' faces and provide support and encouragement as needed. No machine can do that.

## Discussion Questions

1. Many calculations and algorithms that used to be at the heart of math education can now be carried out by machines at the press of a button. How should that impact *what* we teach? and *how* we teach?
2. As laptops become more available in the classroom, more students can access virtual manipulatives. How do those compare with

their physical forebears? Think about convenience, discourse, attitudes, and cognition.
3. Is there a middle ground between "easy enough to learn to do in your head" and "unwieldy enough to be relegated to a machine"? Should we or should we not have a category reserved for "should be done on paper"?
4. Should time in math class be allotted to teaching how to use spreadsheets? To teach programming? Can this be done while still advancing our math education goals?

## Notes

1. desmos.com
2. geogebra.org
3. See for example "Add Till It's Plaid" (mathed.page/attc/plaid) for one such lesson.
4. For an extended version and analysis of this activity, see "Make These Designs" on Henri's website: mathed.page/calculator/make-these
5. We found it on desmos.com, among many other examples, by searching for "desmos art turtle".
6. The use of patty paper in geometry class was pioneered by San Francisco teacher Michael Serra.
7. We know of two brands: Mira and GeoMirror.
8. See for example mathed.page/applets.html
9. Another benefit of software such as GeoGebra is that it orients the geometry curriculum toward notions such as rigidity, incidence, and transformations. These sometimes get short shrift in standard textbooks but they make excellent conceptual glue for other geometric learning, and turn out to be the way geometry is understood and developed at the post-secondary level.
10. While it is probably not suitable for the classroom, Euclidea is an app that offers dozens of construction challenges, from straightforward to very challenging. See www.euclidea.xyz
11. See Henri's virtual manipulatives: mathed.page/manipulatives/index.html#virtual
12. These are applets Henri created using GeoGebra: mathed.page/geoboard/geoboard and mathed.page/geoboard/circle-geoboard
13. This problem is from *Algebra: Themes, Tools, Concepts* (mathed.page/attc). For more spreadsheet-friendly material about iterating functions, see mathed.page/iterating
14. Andrea diSessa, *Changing Minds* (MIT Press, 2000).

15. For more advanced applications, see Abelson and diSessa's *Turtle Geometry* (1981).
16. Wikipedia offers a useful summary of Papert's views: look up "constructionism".
17. scratch.mit.edu
18. snap.berkeley.edu

# 8
# Leading Discussions

How to lead a class discussion is one of the hardest topics to address. It is also one on which there is the most to say. There are only a few fundamentally different ways to spend time in the classroom: the teacher can lecture, the students can do work individually or in groups, or there can be some kind of discussion among the class as a whole. This last kind of time is the most unpredictable. Learning how to manage class discussions is a lifelong journey. Whenever either one of us is called upon as an expert to observe a teacher in training and provide useful suggestions, we ourselves learn from the experience.

Personal styles vary tremendously. We have seen excellent teachers who do things opposite to what we recommend. If you are one of those teachers to whom leading class discussions comes naturally, don't let us talk you out of your best game. For the rest, especially for those with very little experience, we hope to illuminate the landscape and introduce a number of techniques and considerations. If you are planning to try some of these, please keep in mind that no amount of advance preparation or advice can substitute for on-the-job training, the essential ingredients of which are criticism and imitation. This is why we advocate fiercely for collaboration time (for teachers to observe each other and discuss) and in-service time (for teachers to receive ongoing instruction). On the positive side, the techniques in this section are all valuable on their own. It's OK to pick and choose, trying those that make the most sense to you and keeping those that seem to work.

People often use the term *Socratic discussion* for the sort of classroom discussion where students make significant and occasionally unexpected contributions, and where the teacher guides the class to deeper and better articulated ideas. There is another sort of class discussion where teachers explain ideas in a sort of interactive lecture. The teacher does most of the talking, asking occasional questions whose answers they telegraph to the stronger students, who obligingly answer quickly, thus allowing the lecture to proceed apace. A variant of this involves the teacher asking students to complete unfinished sentences. There is certainly a place for this sort of lecture, and it can be an improvement over lectures that involve no student contributions at all.

The reason we concentrate on Socratic discussion is that the other kind rarely leads to deeper understanding. The other kind can be a valuable device to keep interest and attention during a lecture but it does not provide the following ingredients we would expect from a good class discussion.

- Intellectual engagement from a greater number of students.
- A chance for individual students to practice articulation.
- A chance to practice evaluating an argument.
- A chance to be noticed for something meaningful.
- A demonstration that students can come up with original mathematics.
- Reward for initiative.
- Ownership of the final result as a communal endeavor.

Having delineated these, we are going to drop the word "Socratic". For us, the term "discussion" will mean all these things: something beyond token participation by students. We will start by sharing some of the influences that shaped our approach early on in our respective careers, and then go on to spell out some of what we have learned over our decades in the classroom.

## Project SEED

Like most teachers, we learned the craft on the job, with the help of colleagues. Both of us learned a lot from Project SEED in the 1970's. Project SEED was launched and led by William F. Johntz. It was an approach based on whole-class discussion techniques, combined with a trust that all students can engage in abstract and logical thinking if given a chance. We saw that fourth graders in inner-city schools were comfortable exploring negative numbers, exponentiation, and finite groups – as long as this content was taught using the project's discovery methods. This was an ambitious and effective program, which started in Berkeley in 1963, was deployed nationwide, and finally closed shop in 2016, after having helped hundreds of thousands of otherwise underserved students.

Project SEED teachers were mathematicians and graduate students who worked 40 minutes a day in public elementary schools. One day a week, they observed other teachers in action. On some of the other four days, they expected visits and feedback from their colleagues. Interestingly, the feedback was overwhelmingly about what was happening among the students, not so much about what the teacher said and did. Or rather, it was how teacher choices affected student participation.

Henri writes:

> In preparing to write this section, I read through Project SEED's Guidelines for Discovery Teaching, with the intention of sharing some of what I learned some 50 years ago. I was stunned to discover that many, many of my whole-class teaching techniques originated there. They were so ingrained in my classroom demeanor that I forgot where they came from. When I shared these ideas with colleagues over the years, most assumed (as I did!) that I had created these strategies, that they just reflected my personality. In fact, they originated in Project SEED.

We will share some of those here. If you want the whole package, go to the Project SEED page on Henri's website.[1]

## Hearing from All Students

Many teachers, well aware that merely talking at the class is not that effective, try to involve the students by asking questions, and resign themselves to having the same two or three students answer them. Alas, that is not much of an improvement over a straight lecture. It may even be worse, as it gives the impression that students know what's going on – when in reality it is only those two or three students who are actually on board.

Project SEED proposes a number of techniques to get everyone to engage, and (crucially) to let the teacher know where the class stands. This involves varying the student response mode.

- After asking a question, do not tolerate students blurting out the answer. Ask for "quiet hands". Count the number of hands. "… five, six, seven… Seven of you have an answer. I'll give you more time to think". Or else: "talk to your neighbor about this, and I'll ask again". Or: "Who can give a hint to the class? I'm sure more than seven students can figure this out".
- "Show me the answer on your fingers". This works best for small non-negative integers, but students enjoy the search for ways to show $-2$ or $\frac{3}{4}$.
- "Point up if the answer is positive, down if it's negative, make a zero with your fingers if it's neither".
- "Graph the function in the air".
- "Write down the answer" (quickly walk around to see what they wrote).
- If you're pretty sure almost everyone knows the answer, ask for everyone to answer in chorus – perhaps in a whisper.

The younger the class, the more they enjoy these techniques (but don't rule them out for high schoolers!). For students of all ages, they get you out of the trap of relying on the two or three most vocal kids to answer all the questions, a sure way to lose contact with the majority.

In some of these modes students will just imitate other students' answers, since they're visible to all. But the increased involvement overall means it's worth it, and in any case, you can tell if they're doing that, which is itself useful information. Sometimes you can use gestures to tell students to put their hands down as soon as they know you have seen their answer – this makes for a bit of excitement.

### Assorted Discussion-Enhancing Tricks

Even the best questions are ineffective if students are not paying attention. The board is the visual support for the discussion. Here are some tricks to maximize its usefulness, and to get focus.

- Use color, boxes, circles, and arrows for emphasis.
- Erase strategically. "Oh no! I erased the most important thing! Help!" Or else "I need space. I'm about to erase almost everything. What should I *not* erase?"
- Use students' names: "Asa's formula", "Lucy's trick", "Maria's theorem"
- Move around the room.
- If appropriate (e.g., when exploring a pattern), ask "What is my next question?" "Yes! You must be psychic! How did you know?"
- Especially if a student stalls mid-response, or gives an answer that generates many thumbs down: "Pick someone to help you".

If a lesson is going well:

- "Fasten your seatbelts! This is going to be challenging!"
- "This next problem may be too hard. Last year's class couldn't handle it. Maybe I shouldn't even ask that". (Then ask it.)

### Big-Picture Principles

And here are some big-picture principles from Project SEED.

- Embed review in new material, rather than going over old content the same way with promises that it will be useful later. It's more respectful of the students and more interesting for everyone.
- Vary the difficulty, rather than going monotonically from easy to hard. This helps keep everyone engaged.

- It's OK to leave some questions unanswered and return to them in a future class.
- Go back and forth between the general and the specific. Sometimes: "This idea works for small numbers. I wonder if it's always true". And sometimes: "Does this idea work for small numbers? Who has a way to test that?"
- Students should be comfortable making mistakes, as it is often a necessary stage on the road to understanding.

And the biggest principle: if we teach effectively, all students can do math!

**And There's More!**

Just to be clear: There is no one way. We do not claim this is the only way to teach math. But combined with other approaches (such as especially group work) the Project SEED techniques are a powerful addition to any math teacher's toolbox. They add an element of drama to whole-class discussion, and break the hold of the dominant culture ("I explain, you practice".) Coming across Project SEED's Socratic method early in our respective careers had an enormous impact on us. We have used these ideas for about 50 years at all levels, from counting to calculus, and passed on many of these techniques to colleagues and mentees.

**Every Minute Counts**

Henri writes:

> When I started teaching high school, there is one book that I found extremely helpful: *Every Minute Counts*, by David R. Johnson.[2]

In the years since the book's publication, society has changed, math education has changed, and we have changed. Back then, the teacher was at the center of the action in any classroom, including ours. Now, the action is more spread out, thanks to the much greater use of group work. We decided to take another look at Johnson's book, to see how well it's holding up.

Early on, Johnson shares his goals as a teacher. Here are three good ones:

- I will give no options for non-participation.
- I will allow students to make mistakes without fear of failure or embarrassment.
- I will encourage student interaction during a portion of each class period.

After listing his eight goals, he adds: "I don't remember a single day when I can honestly say that I have achieved all of them..." That's evidence that the author is an actual teacher, not someone promoting the latest fad. In fact, the whole book confirms that: it consists of a mix of big picture ideas and specific implementation ideas about everything from homework, to the daily routine, to assessment, all of it stemming from one teacher's experience.

The most useful part of the book, both in the 1980s and now, is the chapter on "The Art of Questioning", in which he presents some great ideas about how to run a class discussion. He starts by challenging what he calls the "one-on-one questioning method" (you ask a question, get the correct answer from a student, and move on, without knowing whether anyone else in the class understands what you're talking about.) He then offers a list of twenty "try-to" principles of questioning. (Yes, twenty!) Here are three of the best:

- Try to avoid yes/no questions.
- Try to ask for reactions to student answers.
- Try to follow up student answers with "why?"

The "why?" question should be addressed to the class as a whole, not to the student who provided that answer. The reason is that if students come to believe that answering a question is sure to lead to a follow-up question, many are less likely to answer in the first place.

Johnson's overriding philosophy of class discussion is two-fold: on the one hand, the questions should get all students to engage, and on the other, student answers should reveal to the teacher the extent of student understanding. He discourages us from asking "Does everyone get this?" or similar questions, which rarely yield accurate feedback about what is really going on. He believes it is the teacher's responsibility to figure out who gets it. This leads to his key insight: he needs answers from every single student. To achieve that, he has the whole class answer questions using the "paper-and-pencil method". He asks a question, has all students write down their answers, and he walks around to see what they wrote. This way he can proceed with the lesson with full knowledge of who's on board, whether he needs to backtrack, and so on. (Of course, there are other ways to get this sort of whole-class feedback, such as the ones we mentioned above in the Project SEED section.)

Even in a collaborative classroom, teacher-led discussions play an important role. David R. Johnson's techniques for genuine student engagement in class discussions are mostly excellent, and yes, they are still valid after all these years.

## Sequencing Discussions

In the next chapter we discuss sequencing the curriculum. Sequencing also applies to a lesson plan. Both of these sequencing activities can be done offline. Our topic here is the conscious sequencing of student discussions. This is much trickier because it occurs in real time and is not completely predictable. This is one area where teachers will find it very helpful to have mathematical preparation going well beyond their classroom's level, as well as a good grasp of pedagogy and time to work on this in a classroom setting collaboratively with other teachers.

Class discussions can arise in different parts of a lesson. Some teachers like to introduce topics in a discussion rather than a lecture. The teacher poses questions so that students can weigh in immediately, without pausing to work on the problem at their desks. More commonly, students are given activities to work on, after which discussions take place about what the students have done. Discussions can also arise after an activity is more or less complete, or spontaneously during a time the teacher is explaining something.

Our discussion of planning overlaps substantially with Smith and Stein's *Five Practices for Orchestrating Productive Mathematical Discussions*.[3] This book assumes that discussion takes place after students have done work individually or in groups; however, most of it also applies to the other contexts for discussion. The book breaks the task of sequencing discussion into five areas:

- anticipating likely student responses to challenging mathematical tasks;
- monitoring students' actual responses to the tasks (while students work on the tasks in pairs or small groups);
- selecting particular students to present their mathematical work during the whole-class discussion;
- sequencing the student responses that will be displayed in a specific order;
- connecting different students' responses and connecting the responses to key mathematical ideas.

Smith and Stein recommend a process of anticipating that is a little more detailed than we would recommend: they suggest solving the problems multiple ways and getting colleagues to do so as well in case they find ways that you did not. Our experience is that anticipating student responses is more a matter of experience and content knowledge than of divination. If you have taught this lesson before and been surprised by what some students did, make a note of it for next time. Think it through if you were not able to figure out in the classroom what this student was doing. This might be a time to prevail on your colleagues for help. If you keep your lesson plans on your computer, make a second copy, to which you add post-lesson notes, which will be the basis for next year's (or next period's) lesson plan.

Having enough content knowledge to grasp the connections between this lesson and everything else the students have learned will help you to anticipate how students will think. But even the most experienced and knowledgeable teachers can count on a regular stream of surprises. Over time, one gets better at adjusting sequencing on the fly.

Monitoring refers specifically to the model in which students work individually or in groups before the discussion. During this time, often the teacher is busy with multiple tasks, one being to help students who may be stuck. When discussing cooperative learning, we mentioned the need for the teacher to keep moving. If each group seems to be requiring a sit-down with the teacher, then probably the task was too difficult or too vague and should be restructured before it is used again. Assuming the teacher is able to circulate efficiently, they will be able to make a mental note of roughly which group has done what. One advantage of students working on whiteboards is that the teacher is able to monitor the groups from a distance.

Selecting students to present will be discussed in subsequent sections because of the psychological and social component. It is usually not feasible (or helpful) to hear from every student every time. Once you have chosen a small number of students to present their solutions, sequencing is very important. A good general principle is to have students present in order from least complete to most complete solution. When all the ideas in a student's solution have already been stated, it is anti-climactic for the student and boring for the rest of the class. By contrast, when each presenter has something more than what was said before, the whole class remains more engaged and the intrinsic reward for presenting is greater. The same result can be achieved by asking "who can share their first step", and then "who can help us take the next step?", and so on.

Beginning with less sophisticated solutions is also a good idea. For one thing, students have the best chance of following more sophisticated solutions after they have mulled over the simpler ones. For another, this often reflects the progression of students' individual thought processes. Thirdly, when discussing solutions, the less sophisticated approaches often serve as a framework on which the more sophisticated solutions build. Often the less sophisticated solutions are less complete as well, and these two sequencing criteria coincide.

Smith and Stein make a point to distinguish between class discussions that build to a greater common understanding, from what they call show-and-tell sessions, with different groups presenting their solutions in a disconnected fashion. In their introduction, they say,

> In fact, we argue that much of the discussion in Mr. Crane's classroom was show-and-tell, in which students with correct answers each take turns sharing their solution strategies. The teacher did little filtering of the mathematical ideas that each strategy helped to illustrate, nor did he make any attempt to highlight those ideas. In addition, the teacher did not draw connections among different solution methods or tie them to important disciplinary methods or mathematical ideas. Finally, he gave no attention to weighing which strategies might be most useful, efficient, accurate, and so on, in particular circumstances.

To get the most out of student presentations, incorrect or incomplete solutions must be critiqued. There is a lot to learn from where and why a solution went off the tracks. This does not have to be a negative experience for the students being critiqued. Many, perhaps most solutions emerge imperfect. The students are capable of understanding this when it is a part of the daily routine. Sequencing from least to most sophisticated allows the class to appreciate the positive contributions of the cruder solution attempts. Still, critiquing student ideas needs to be done with some sensitivity. We discuss some effective approaches below, under "Handling Wrong Answers".

If different groups or different students present substantially different approaches to a problem, it can be illuminating to look for connections between them. For example, how does this algebraic solution relate to this guess-and-check approach? or to this graphical method?

In *Principles to Actions*[4] as well as *Five Practices*, a distinction is made between criticism supplied by the teacher and student discussion of solutions. Both books give many examples of ways the teacher can guide students to express ideas, compare them, locate problematic areas, and repair them. When students learn to do this, they accelerate their capacity to learn, and bad teaching later often will not stop them!

Related to the comparison of solutions is the final step of ensuring connections. This encompasses some of what we just said concerning critiques of solutions, as well as something we discussed in Chapter 1, namely the institutionalization of knowledge.

A number of very detailed examples of these principles for sequencing discussions are given in *Five Practices*. Here is one example we like from *Principles to Actions*. The students are given the following problem:

> At a baseball game in April, a batter came to the plate, and the scoreboard showed that he had a batting average of .132. After

two pitches, he hit a single into center field, and the score-board immediately updated his average to .154. Given this information, determine how many at-bats he has had so far this season, and how many hits he has had.

This is a case where some advance planning will help. Is there only one solution or are there multiple possibilities? When a major feature of the problem is not obvious, it's best to work everything out beforehand.

We can identify several strategies for solving this. One is by trial and error. There are several ways to search for a pair of whole numbers whose ratio is 0.132. Suppose the kids have calculators. (Planning step #1: realize that you would not want to do this problem without a calculator.) Start with a guess. Then keep increasing either the numerator or the denominator depending on whether the result is too big or too small. Another way is with a spreadsheet. Look at every number (the number of at bats) multiplied by 0.132 until you get something close to a whole number (the number of hits).

A more sophisticated strategy involves algebra. If $x$ is the number of hits and $y$ is the number of at bats, the two pieces of data give two linear equations which can then be solved. There are pitfalls. The variables need to be described better than in the previous sentence (number of hits and at bats at what point in the story?). Be on the lookout for this when monitoring student responses. In our experience, students, if they use algebra at all, don't articulate the variables and end up not being able to use both pieces of information.

Another pitfall in the algebraic solutions is that when the student correctly formulates and solves the problem, $x$ and $y$ will not be whole numbers. Then what? Both methods above, and any others we have not mentioned, lead to the same obstacle: it will be necessary to identify how a scoreboard rounds off. You may have to provide this knowledge. This results in inequalities rather than equations. A complete answer involves finding all possible solutions and arguing that there can't be any others.

It will be worth the time in advance to ponder how to lead a discussion that brings out the successes of some students' trial-and-error approaches. The discussion will need also to bring out some criticisms: is it sufficiently obvious how you do this when the numbers are different? Can it be done systematically enough always to lead to a solution? How can a trial-and-error approach ever be sure it has found all solutions? (It often can't – that is a common weakness of trial-and-error approaches.)

The algebra discussion might be organized along obvious lines, beginning with a group that had the idea to use algebra but trouble formulating equations, to one that formulated, to another that formulated cleanly. The discussion of mis-formulated variables could be productive if you give it a little time. The next student to call on after this would be one with a solution to the algebra problem. Even if one of the groups you called on for a formulation had a solution, it may be best to spread the mic around by letting another group contribute the solution. It will be obvious that there is an issue with their not being whole numbers. This is a good time to ask the students their ideas on a way forward. Ideally, they will see on their own that they need to know how a scoreboard handles decimals. You should be prepared, though, to nudge them.

Also, you should be aware that a complete argument that all the solutions have been found is probably a little over-ambitious for a typical class that would be discussing this (it is Algebra I material). It's perfectly fine to stop short of that, but also good for the students to identify that this could and should be the goal. That might be the main take-home point of the lesson! If so, make sure it is institutionalized.

## Staging

We begin our discussion of on-the-spot techniques with some considerations that may seem trivial, akin to beginning rehearsals for a play by talking about the blocking. However, you will see that most of our suggestions are as much about the psychology as the mathematics. A teacher's expectations of a class discussion are conveyed nonverbally as much or more than verbally, and physical aspects such as stance, eye contact, and tone of voice play an undersung role.

Inclusiveness increases when you put as much of the class as possible between you and the respondent. If you call on a student on the left side of the room, walk over to the right side as you're doing so. As the words flow between you and the respondent, the almost physical presence of a stream flowing between the two of you will wash over the students in between. When you and the respondent look at each other, some measure of eye contact automatically extends to the rest of the class.

It is often a good idea to get students to come up to the board. Students will give longer monologues at the board than they will from their desks, so be prepared to be a more active moderator if a student is going on and on or has lost the rest of the class. Having the right solution should not be a prerequisite for coming to the board. A student explanation is not a surrogate for a teacher explanation, it is the beginning of some kind of back and forth. For this to work, volunteers must not be prescreened for correctness. There is some tension between this and two other goals: efficiency and encouragement. We will have more to say shortly about keeping the lesson on track and helping students avoid embarrassment.

Related to this is some important body language. When a student is at the board, try to take up a position in the back or on the perimeter of the room. Sometimes one can simply sit in the student's seat. This has the effect of including the rest of the class while also giving you – the teacher – a student vantage. You may be surprised at what you see. Other body language to be aware of is whether you are passing judgment on what you hear. Do your eyes flit impatiently with wrong answers? Do you gesture in agreement with right answers? Do you angle your body to the board as if to write down something correct, then pause if it's not what you wanted? None of these is a killer, in fact we all do these things, but tip-offs like these do work against students developing their own judgment.

The subject of tipoffs leads to the topic of intentional errors. These are a hit with younger kids, in a slapstick sort of way, but they lead to earnest engagement in all age groups. If you are going to use these, it's best to have a plan from day one. Start with some obvious ones (e.g., a mistake in basic arithmetic). Continue doing these many times per lesson, until the kids become used to raising their hand to correct you. The goal is that students are always listening critically, and begin to apply this to more substantive instances. Older students can feel patronized, so the mistakes need not be as blatant. For them, you can announce "What, no one noticed my mistake?", or "Luckily, I never make mistakes, so I'm going to do this quickly". In a course you have taught several times, you already know what mistakes are likely, so you should make sure to get in front of this by making the mistake yourself early on.

## Encouraging Without Babying

We cannot recommend too highly the book *How Children Fail* by John Holt.[5] Holt's journal entries, recounted in the book, observing math classes as a part of his graduate work document the psychological mine-field of early grade education. Among his observations are that classroom dialogue is guided by many principles. One is avoiding embarrassment. Another is getting or avoiding attention. A third is to please the teacher. Most of these are stronger, at most times, than the desire to seek the truth.

The tension between soft and hard teaching comes to a head here. How can a teacher deal with wrong answers and ideas with intellectual honesty and yet protect the delicate psyche of the students? One important technique is, when possible, to recognize what is right with an answer that is, for the most part, wrong. Answers usually don't come from outer space. If a student says that $x^2 - 4x$ has a root at 2, you might say, "yes, 2 is a root of $x^2 - 4$, but what about when the second term has that $x$ there?" It does not really matter if you mis-guessed the path at which the student arrived there. What matters is that some sanity was granted to the answer, the flaw was pointed out, and the question was put out there again.

What if the student had said not 2 but something baffling like 7. You might try, "Thanks for giving us a number to try out! Looks promising, because $7^2$ is bigger than 7 but then when we subtract $4 \cdot 7$ that might be enough to bring it back to zero. Let's give it a shot: $7^2 - 4 \cdot 7 =?$" You might even get a payoff from 7 as a generic guess, serving the way $x$ would: another student (or the teacher if necessary) can compute this by first simplifying to $7 \cdot (7 - 4)$. No, it does not come out to be zero, but maybe it's more obvious now what would.

A device we commonly use to encourage participation is at some point to ask for guesses. We do so when most students are not in a position to get the right answer. You may attribute the guesses by writing a student's initials next to each one on the board. If there are a lot of guesses, it's like playing the lottery: excitement for a student who wins, but no shame in being one of many who did not. Answers need not always be attributed. You can even act like you're thinking, and add your own incorrect guesses. The key is to put some distance between the wrong answers and the students who made them.

When you expect only a couple of answers, write the first answer you get on the board without attribution and ask the student to choose whom to call on next. They immediately bask in the power that has been handed to them and the mistake is no longer important. You can exert some control over the guess process. If you want to get a variety of guesses, avoid calling first on a known "smart student" that everyone will assume has the right answer. Even if he or she does not, kids will be reluctant to contradict the answer. Another technique is to ask everyone to write down a guess.

Sequencing students' answers so that later answers build on earlier ones is encouraging. One or more solutions will be accepted by the class, perhaps having been refined a number of times. Contributors in the early stages feel they retain a portion of the credit. This will be true even when the solution that builds on theirs also changes it, even corrects it. In the wrap-up, being explicit about how one solution subsumes another will help the students make connections. Thus, the summary, "Can we agree that Hiroshi's idea to identify the rate of square feet of painting per day does what Jack did, but in a more abbreviated way?" not only shows the connection between the repeated addition interpretation of a rate and the more sophisticated proportional interpretation, but it also gives Jack the encouragement he deserves for finding a solution from scratch.

Finally, classroom discussion is a great vehicle for improving students' verbal math skills. Emphasize that the hardest thing about math writing is coming up with a first draft. Keep in mind that the first phrasing will be awkward, nothing like what a teacher would say. A little editing on the board by their peers will produce a collaborative result that the student can be proud of. On the other hand, being corrected by the teacher can be taken as a putdown, and discourage further contributions. (On the other hand, if the statement is out-and-out useless, it's not a good time to make an example of the poor wording on top of that.)

## Handling Wrong Answers

The message to the students has to be that it's OK to make mistakes and that in fact, it is the only way to learn math. Some teachers are very nervous about ever telling a student their answer is wrong. This is of course problematic, as it makes it very difficult to conduct a discussion, and all this tension only increases student anxiety. We should strive instead for a classroom where incorrect answers are voiced frequently, and discussed as learning opportunities, thus making it possible to discuss mistakes openly when necessary.

How to do this is complicated and there is no single correct answer for all situations. Let us first clarify our goals:

- broad participation by students in the conversation;
- progress toward better understanding for most;
- correctness determined by discourse, not by authority.

Teachers often complain that it is always the same few students who raise their hands in classroom discussions. There are many possible reasons for this, such as not giving students enough time to think, not letting them practice their answer by talking with their neighbors, not asking the right questions, and so on. But one huge reason is students' fear of being wrong. If a wrong answer is met by ridicule from classmates or teacher, that is sure to cut down on participation. But this sort of intimidation can happen in more subtle ways: if we think of classroom discussion as a way to quickly get to the right answer, heap praise on students who supply correct answers, and move on, the message to students is that they'd better not speak up if they are not totally confident about their answer.

So, it is important to not rush to the punch line. If something is important and difficult, there are probably misconceptions in the class, and rushing to the right answer will allow those to remain unchallenged. We should strive for an atmosphere where wrong answers are expected, and in fact appreciated. The classroom culture has to make it comfortable to be wrong, as it is really the only way to learn to be right. It is tempting to praise correct answers, but in the long run it can have a chilling effect, as students will hold back from speaking up until they are sure they have a correct answer. For some students, that certainty never comes. Correct answers are their own rewards, while a brave attempt to answer a difficult question, even incorrectly, needs explicit support from the teacher, who should instead praise and encourage participation and risk-taking. "Maya was brave, and raised her hand even though she wasn't sure her answer was right. Thank you, Maya!"

This does not mean that you should *never* acknowledge correct answers. It's more a matter of how. Scott Farrand writes:

> If a student, especially one who does not often share their ideas or who has a low opinion of their mathematical abilities comes up with a right answer, I probably will provide a direct opportunity for their fellow students to express praise, by asking something like, "How many of you think that is the idea that we really needed?" Some students have built their identity around their skills in a video game, or the clothes they wear, or all sorts of superficial things. They are unlikely to change that because their math teacher praised an answer they offered. They usually hold those values because it garners some esteem from their peers. For the student who is too cool for the room and rarely engages, there is nothing more powerful than having the whole class give them positive reinforcement for a really good idea.

Still, there is a sting when one gets it wrong. One way to diffuse this is to ask the student who made a mistake to choose a classmate to help sort things out. This shifts their focus to this newly acquired power. Another approach is to routinely ask for many answers, whether the first answer given is right or wrong, write them all down, and discuss how one would sort out which one is right, perhaps after voting on them. Teacher mistakes (made on purpose, or not!) should be a frequent feature of class discussion, and being relaxed about them helps create the right atmosphere.

Many of these strategies rely on keeping a poker face and using classroom discourse to address the errors. These strategies do support the three goals mentioned above. However, it is possible to overdo this and to never ever make clear that an answer is wrong, instead falling into an awkward silence for fear of hurting student feelings. This does not work: students can read our body language in those situations, and in any case, will sooner or later realize they were wrong. If you don't have a strategy to handle the situation, frankly, it's better to out-and-out acknowledge the answer was wrong, and thank the student for offering it. "Thank you Charlie! Other students almost certainly thought that was the answer, and not discussing it would not help, would it!"

## Listening

Some things almost go without saying, having to do with maintaining a respectful environment. In classroom dialogue, students venture ideas, voice disagreements, and grapple with each other's contributions. For students to participate at that level, teachers have to make sure it is a safe environment. They cannot tolerate students making fun of each other, even if they're "just joking", and they themselves should refrain from sarcasm.

But listening is more than being respectfully quiet. It is about taking the speaker's ideas into your head and engaging with them. One barrier to listening is the need to formulate a response. If six kids are wildly waving their hands in an attempt to be called on, when you do call on one, you can bet the other five will be busier rehearsing their own answer in their head, in case they are called on next, rather than listening to what the first student has to say.

Another barrier is interruption. Math educator Tom Lester once mentioned a study showing that the average amount of time between when a teacher asks a student a question and when the teacher prompts the student or gives up on them is 2 seconds. Two seconds is longer than it sounds, but nowhere near long enough to formulate a coherent thought unless you were already thinking it before the question was asked.

There may not be anything you can do about the sound-bite trend in news reporting, but there's a lot you can do about it in your classroom. The first thing to try is waiting. Don't nod yes or no, or say uh huh, or give the student any feedback at all until they have finished saying what they wanted to say. Then wait five or ten more seconds. The odds are that the student will, after pausing for breath, realize that they are not finished and continue. If not, at least the other students will have had a chance to think about what they just heard. If you're uncomfortable with such a long pause, try pacing back and forth or holding eye contact with the respondent as if you expect them to continue, or act as if you're trying to digest what they've just told you. In fact, often you really will need time to think. If they said something that was wrong in a puzzling way, see if you can figure out what they really meant. Students will only listen to each other if you set an example, so make sure you don't respond without having really heard.

In a class of 20–30 kids, it is easy for a student to tune out, take a few notes, and plan to sort it out later. Students may also feel they have no right to question because everyone else obviously understands. You can counter this by demonstrating your expectation that each student must understand what each other student has said. After one student says something the slightest bit unclear, ask another to repeat it in their own words (thus practicing precision, a meta-curricular goal). This is a good time to pick students rather than have them raise their hand to volunteer a paraphrase. If student B can't paraphrase what student A said, it's not necessarily student B's fault. Student B can ask student A to clarify if necessary, or ask for volunteers for someone else to clarify. Make sure to go back and find out whether student C's clarification of student A's remark did in fact help student B. After a little experience you'll know better when to go through this routine. If student B simply was not listening, they might feel reprimanded, but that's OK. The basic standard you are setting is that the discussion involves the whole class and is not a collection of one-on-one dialogues between the teacher and individual students.

We summarize with a short list of techniques that are taken from *Five Practices* but might equally well have been lifted from the Project SEED manual 50 years prior.

- Using wait time.
- Asking students to restate someone else's reasoning.
- Asking students to apply their own reasoning to someone else's reasoning.
- Prompting students for further participation.

## Questioning

Varying the level of difficulty of your questions is important. Don't worry too much about a question being too hard or too easy. A hard question can always be followed by a hint. You want to stay away from questions that ask the students to read your mind. Questions that are less about the mathematics and more about what the teacher had in mind are not easily followed up with hints and it can be absurd to try. A question like "What might be a way to get $x$ on the left side of the equation?" or "Do you see an expression that can be factored out?" pre-supposes the student wants to follow the same path the teacher has in mind. It is better to ask a broader question: "What might we do next?" "Does the equation simplify at all?" "Should we use trial and error?"

It is also a mistake always to use questions that are too easy in order to draw out weaker students. For them it is better to use questions that are not right/wrong, such as what would be an interesting number to plug in, which of the previous remarks do you agree with, what did you think was the hardest question on the worksheet, or the most interesting? The danger in using a too-easy question to draw out a timid student is that it is a no-win situation. At best the student says the right answer, which is not too impressive. Everyone knows you chose this student for a softball question. At worst the student flubs the easy question, which is doubly embarrassing. You might be wrong about the student: students can be quiet because they are bored. A better strategy is to make the reward higher. Long-term confidence is built by the realization that the student can answer something nontrivial.

On the other hand, there is a place for easy questions, for example, to remind the class of yesterday's discovery, or just to make clear what we're talking about today. They need not be directed at a weaker student, or at any student, and anyone can be chosen to answer. In this situation, one useful format is to ask the question, and then ask the students to tell the answer to the person sitting next to them. In addition to providing participation for everyone, this prepares students to volunteer an answer for the whole class.

Some questions can be planned, but most are on-the-spot inventions. Inexperienced teachers can freeze up. Experienced ones can get in ruts. For this reason, it is useful every now and then to look over a collection of offbeat questions, such as this list contributed to an online forum by James Propp and Jim Tanton.

- Does this answer make sense?
- Is there another way we could arrive at this answer?
- Is there a pattern here?
- What mistake did I just make?
- How am I fooling you?
- Are we using the right definition?
- Have I given you enough information to answer the question?
- What other information might you need?
- How convinced are you?
- In plain English, what is this equation telling us?
- Now that we see what the answer is, could we have seen that more swiftly?
- Why would anyone want to answer this question?
- Oh bother. I don't see what we need in order to proceed. I think we should weep. (To spur the class on to do something with the issue/question at hand)

A section heading in NCTM's *Principles to Actions* is "Pose purposeful questions". Another good way to broaden your repertoire of questions is to pick a purpose you feel you have been ignoring and try for a week to emphasize questions that accomplish it. NCTM lists four broad purposes in that section.

- Gathering information.
- Probing thinking.
- Making the mathematics visible.
- Encouraging reflection and justification.

If you are interested in pursuing this, it is worth stopping now and reading that section, which includes a long example scenario and an analysis of it.[6]

## Answering Questions with Questions

When you challenge students to think, they will find creative ways to get you to provide the answers. One is to start to guess while waffling and trying to read your tip-offs. Another is to come back at you with a question before they have really tried. Experienced leaders of Socratic discussions often develop reputations for answering questions with questions.

Regardless of whether you think this is a good or bad reputation to have, there are certainly better or worse ways to do it. Generic answers sounding like non-directional therapy are covers for a brush-off which students will not appreciate. The lesson will move forward much better if you find a meaningful response, offering some direction without handing over more of a hint than you want to. Here are a few examples of generic responses and more meaningful alternatives.

**Student:** Is this right?
**Instructor (generic):** I don't know. Does it sound right to you? Can you elaborate?
**Instructor (responsive):** Are you asking whether your computation is correct, or whether it will prove useful?

**Student:** What should we do from here?
**Instructor (generic):** What do you think? Does anyone have any ideas?
**Instructor (responsive):** Why don't these equations tell you what $x$ is?

**Student:** Can we set $z$ to be the average?
**Instructor (generic):** I don't know; can you?
**Instructor (responsive):** If you're wondering whether you can assign the variable $z$ to be the average of all the prices, the answer is yes, but you haven't yet said whether we know anything about $z$.

## Obtaining Mass Response to Questions

Some college campuses use clickers or phone apps to allow a whole class of students to register an answer. For younger students, a low-tech alternative works perfectly: hand signals. Project SEED used agree and disagree hand signals successfully for decades. A student registers agreement by pumping a fist, and disagreement by moving hands back and forth, palms down, crossing then uncrossing, like a football referee indicating an incomplete pass. We demonstrate these gestures in Figure 8.1.

**Figure 8.1** Project SEED's "agree" and "disagree" signals.

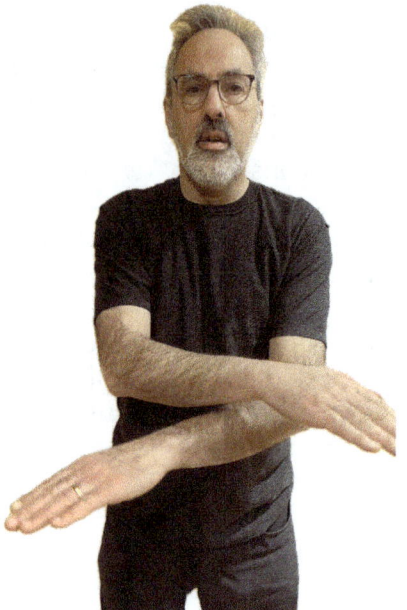

The agree and disagree signals are arbitrary; you can pick others if you prefer. Some teachers like thumbs up/down/sideways for agree/disagree/don't know, perhaps encouraging the students to hold the thumb close to their chest, to minimize copycat responses. In any case, it may take time for the hand signals to take root. You should introduce them early and request them daily. It is best to ask for them in response to a question, and to avoid their use while a student is struggling to formulate an answer.

**Keeping It Interesting**

You can try to keep classroom discussion varied by using humor and stories, by physically moving around the room, and once in a while by teaching in total silence. (Ask questions by writing them on the board. Hand the marker or chalk to a student so they can answer without speaking.)

You can vary the ways you display ideas: on the board, projecting a computer screen, or using manipulatives on a document camera.

Finally, instead of "going over" a problem, it is a good strategy to use a problem that is slightly different, so that the students who successfully completed it in homework or class work have some reason to pay attention.

## Cold Calling

*Cold calling* in education refers to asking a particular student to answer rather than selecting from among students who volunteer, say by raising their hands. Some teachers never do this. For others, it is a matter of course. The end goal is for everyone, not just the extroverts and show-offs, to experience the learning that comes with articulating their thoughts, being part of the conversation, and influencing the flow of ideas. The downside is that a shy student, when put on the spot, can react with fear and become even less likely to participate in the future.

Robin writes:

> I don't think one should make any hard and fast rules about this, but here's what I do. I try to habituate the students to being called on right from the start, but only conversationally. What did you think of the homework? Were you able to hear what Bradley said? Which problem is your group on? Can you give me an input for my function? Even these questions cause a bit of discomfort for acutely shy students but it's necessary for them to get past this and this is the gentlest way I know to loosen them up. Later in the semester, questions like these might indicate that I think the student can't handle anything harder, but at this point I don't know who is confident and who is not, and the students realize this. Pretty soon I can ask more substantial questions: not "Did you hear?" but "Did you understand, and if so, do you agree?" (And if you did not understand, let's get some clarification and try again.) Again, I'm stepping up the cold calling content for everybody, not just for the shy ones. I would like it to be no big deal when the cold calling begins to involve real math. After a few weeks I become aware of who is not volunteering and I make special efforts to reel them in. I do this because I think one of the most important things for a teacher to achieve is inclusion. Any student who is not participating is not having their needs met. It is my job to reverse this. I have to do it carefully so I don't make things worse, which is why I don't have hard and fast rules.

When making your decisions about how hard to push student engagement, consider the reasons for achieving it. One is that part of what they are learning is experiential and is not the same if they are watching someone else do it. Another has to do with whole-class discussions. Students need to understand that they are speaking to each other, not having a sequence of private conversations with the teacher. This is not credible if the other students cannot be counted on to respond. A third reason is to increase every student's confidence that they can generate mathematical ideas on their own. In fact, all seven benefits of Socratic discussion listed near the beginning of this chapter are forfeited by a student who never participates.

Probably the best way to make progress in increasing student engagement in whole-class discussion is to set up a buddy system with a colleague: you will visit each other's classes, and offer feedback to each other with a special focus on student participation. Who engages? Who does not? As the observer, pay attention to how teacher decisions affect student involvement.

## Where We Are

In Chapters 1–3 of this book, we presented an overall pedagogical perspective: do not fall for over-simplified panaceas or dogmatic creeds; put student intellectual engagement at the center; be flexible and open as you implement various ways to achieve that goal.

In Chapters 4–8, we zeroed in on the specific ingredients of day-to-day classroom practice, focusing on what the students do, what tools they use, and crucially how the teacher leads and supports them.

It is now time to zoom out and put all these ingredients in a broader context.

## Discussion Questions

1. A false dichotomy contrasts the teacher's role as "sage on the stage" versus "guide on the side". Yet those two roles are both essential. What are some effective ways to combine them?
2. A student offers an explanation or solution. You can either tell them they were right, or you can ask another student whether they believe the explanation. For each of these two choices, list possible ways it could lead to a better result than the other.
3. Traditional pedagogy emphasizes teacher lectures. Students are to listen carefully, take notes, and memorize. Is there a place for those ingredients in a class where students engage intellectually rather than just clerically? How does the teacher get the most impact on student learning from standing in front of the class?

## Notes

1. mathed.page/project-seed
2. D. R. Johnson, *Every Minute Counts.* Dale Seymour Publications, 1982. The book includes great illustrations by cartoonist John Johnson.

3. M. Smith and M. Stein. *Five Practices for Orchestrating Productive Mathematics Discussions*. NCTM, Reston, Virginia, 2011.
4. National Council of Teachers of Mathematics, *Principles to Actions*. NCTM, Reston, Virginia, 2014.
5. J. Holt. *How Children Fail*. Pitman Publishing Company, New York, first edition, 1964, 1982.
6. *Principles to Actions*, p. 35.

# Part III
# The Big Picture

# 9
# Extending Exposure

Up to this point, we have focused on the daily goings-on in the classroom: different modes and different tools. But of course, each lesson is part of a chapter or a unit. Each unit is part of a course. How should the course be paced? How should the content be sequenced? In this chapter and the next two, we look at the big picture.

Teachers are often given a terrible piece of advice: "Aim for the middle". Every class includes a wide range of talents and backgrounds. Since that is difficult to manage, the idea of aiming for the middle seems like common sense: if you aim too low, you are betraying your stronger students; if you aim too high, many kids will be frustrated. Unfortunately, as is often the case, common sense is not a good guide.

Even in the unrealistic circumstance where all students are at roughly the same place, decisions must be made daily about how much to review, how rapidly to push forward, how thoroughly each topic must be addressed, and so forth. This is the pacing of the course. And many of these decisions are related to the overall sequencing of the curriculum.

All classes are heterogeneous, some more than others. To address this reality, we propose pacing and sequencing strategies based on the idea of *differentiation by time*, not content. We will spell those out, but first, we share some philosophical guidelines.

## Reaching the Full Range

It is important to have some form of support available outside of class for students who need it. How to do this will depend on the specifics of your school. As department chair, Henri asked his colleagues to take turns staffing "Math Café", a classroom where students could bring their lunch and their questions, and get help from a teacher (and sometimes from peers). Institutionalizing that sort of structure makes it easier for students who need help to access it, and reduces any stigma associated with this.

It also makes things more equitable among the teachers: the ones who students feel are most approachable should not bear the main burden of outside-of-class support.

So yes, it is necessary to support the students who need more help. What is not as obvious is the crucial importance of building an alliance with the strongest students. They are the key to our success in a heterogeneous class. This may be the most essential component of a winning approach, and is probably the most difficult to understand. It is predicated on realizing that in a student-centered classroom, the strongest students are the engine that drives the class. This should affect our big-picture planning: if we do not keep the course challenging and interesting to them, we lose their respect and their cooperation, and thus we lose our key classroom allies.

Many of our suggestions are predicated on acknowledging the fact that strong students exist. The reality is of course more complicated than a one-dimensional ranking of students allows. Some students' talents are more quantitative, algebraic, or geometric. (The same is true of teachers!) Some students have a better disposition toward protracted work on a tough problem, a quality that is important to mathematicians but is not necessarily reflected in good test scores. Some students have deep insights but make small errors, and others are fast and accurate but show little ability when working on non-standard problems. Some students thrive under time pressure and competition, and others prefer a more collegial atmosphere. And so on.

Nevertheless, there is a socially accepted view that students who pick up mathematical ideas faster and get good grades are strong students. We would like to argue that these students, and the ones who are strong students when measured against any of the above criteria, are especially important. Here is why:

- *School and district politics*: Strong students' parents are often very active in school politics, so they have the ear of administrators. If their children are not appropriately challenged, they will do everything they can to remedy that situation. Their ideas may not always in fact be in the interest of their children, or of the community as a whole, but they will not tire until they get their way.
- *Pedagogy*: To better support all students, the teacher needs an alliance with the strong students. If the latter are being appropriately challenged, they will contribute to class discussions, help their classmates in group work, and generally have a positive disposition. If not, they will resist, or merely endure the class. This is not conducive to a healthy community of learners.
- *Philosophy*: The most fundamental argument about the importance of strong students is ethical. They are our students and deserve the best education we can give them. Yes, that is true of all

students, but we have a responsibility to society to train the next generation of mathematicians, scientists, and engineers. And the next generation of math teachers!

None of this is to denigrate the importance of all the other students. Quite the opposite: it is to everyone's advantage to take good care of the strong students. And here is an important final point about strong students: the pace of forward motion should be set largely in response to them. This makes it possible to cover material adequately and gives the whole class something to strive for. The use of "semi-optional" extra-challenging material mentioned in Chapter 4 should be part of the picture: by giving speed demons ways to delve deeper, it helps moderate their drive to rush forward to ever new topics. The "eternal review" policies discussed throughout this chapter and the next two will help guarantee this works for everyone.

## Guiding Principles

Classroom heterogeneity is the fact that (as everyone knows) students pick up new ideas at different rates. Many of the strategies we have already shared in this book are about that: using rich activities, the importance of student collaboration (group work), a tool-rich pedagogy, multiple representations, and classroom routines and discussions that involve everyone.

Making progress on all those fronts is essential, but it is challenging, and it takes time to grow in all these directions. Also, it is not sufficient: even with great teaching that incorporates all these ideas, some students just need more time. You may be thinking: there's a lot to cover, and if you had to wait until everyone got it, you'd never get anywhere, and moreover, who wants to penalize the stronger students for the benefit of the others, who may not get it no matter what we do? We've got to keep up the forward motion!

We maintain that you can have plenty of forward motion, while at the same time giving enough time to students who need that. The way to do it is to use some strategies to *extend student exposure* to concepts. The idea is to strive simultaneously for *constant forward motion*, and *eternal review*.

Forward motion is essential to keep a course interesting, especially to our strongest students. Review is essential for all students if we want ideas and techniques to stick. To achieve forward motion, it is necessary to start on a new topic before the previous one is grasped by everyone. Eternal review is of course impossible, but much review can be built into a course without harming the forward motion. This is aspirational, as it is sometimes necessary to pause the forward motion, and there may not be sufficient time or resources for eternal review. Still, it is the right thing to aspire to.

Extending student exposure to the most important concepts and techniques is needed for two main reasons. For one thing, retention is drastically improved if students see important ideas more than once. Also, it is hard to learn when under time pressure, whether facing the tyranny of the clock or the tyranny of the calendar. "You must solve this by 10:15! You must understand this by Friday!" For many students, far from being motivational, artificial deadlines are anxiety-provoking and paralyzing.[1]

As it turns out, it is not difficult to extend exposure without taking more time, by rearranging the time you are already using. We will share practical ways to do this in this chapter, and in the next two. These strategies provide a lot of bang for your effort, and they are compatible with a wide range of teaching styles and curricula.

## Lagging Homework

The first strategy in our exposure-extending approach is *lagging homework*. The idea is to completely separate tonight's homework from today's class work, by assigning homework about ideas that were introduced about a week prior (more or less.) Here are some arguments in favor of this practice:

- Students do not rush through classwork in order to get to the homework and try to do as much of it as possible in class. As a result, they are more intellectually present and more available for reflection, discussion, and collaboration.
- Teachers do not rush through the introduction of a new topic. They are not under as much pressure, because they don't need to reach all students before the end of the period. If today's lesson does not go well, there is always tomorrow. Under this system, if something turns out to be difficult and require more time, there is no need for the teacher to change the homework assignment frantically during the last minutes of class.
- Lagging homework extends students' exposure to the ideas: what could be done in one week (with exceptional students) now takes two weeks, which gives the students who need it more time to absorb the ideas. And this without harming the students who don't need the extra time. In fact, it gives them an opportunity to later review ideas that they may have absorbed too fast.
- This policy is also helpful to the stronger students in another way: it allows forward motion to a new topic before every single last student is ready to move on. The student who is not quite ready knows that they will have another chance to grapple with the idea

in next week's homework. It is differentiation by time and does not require the teacher to come up with a different curriculum for different students.

But, you ask, *are students confused by this practice*? It is probably different from what they've done before, yes, but it does not take long for them to buy into this system, for all the reasons given above.

*Don't students need to practice a new idea soon after they hear about it*? Well, yes, but that need not be done at home. Giving them a chance to do their first practice in class means that they are doing it in the presence of classmates who can help, and of course a teacher. Thanks to better understanding, there may be fewer questions the next day resulting in a better use of precious class time. We recommend a one-week lag (or so), but one teacher told us that even a one-day lag is helpful. On the flip side, exposure can be extended even more by waiting another week (after homework is due) to quiz on a topic, and yet another week before quiz corrections are due. Combining all these techniques can extend exposure time from one week to four!

*How does it work at the beginning and end of the term*? At the beginning of the term, you might give homework on a topic from last term, instead of spending precious class time on it. At the end, you might schedule a buffer week for review. But in any case, this is a general guideline, which obviously need not and cannot be implemented in a rigid manner.

*What if you like to assign homework that prepares the students for the next day's lesson*? We love that idea! It is not at all in conflict with lagging homework. The main point is that on most days, you should not assign homework based on the day's lesson. A week's delay, more or less, provides many advantages, the main one being extended exposure to each topic. This in no way precludes homework that sets up the next day's lesson, as long as it is not based on today's lesson. Such homework can reveal a key idea, generate curiosity, or in some cases dispose of a necessary digression up front. In general, such activities (whether assigned as homework, or as warm-ups) are essentially long-lagged work, based on ideas that were introduced the previous semester, the previous year, or whenever. Such long lags can also be used for review.

On the other hand, if preparing for tomorrow's lesson requires completing homework about today's lesson, then we strongly discourage that because the collateral damage on some of your students would be substantial.

## No Homework?

How about abolishing homework altogether? We sympathize with this question. Most learning happens in class, and one should not overdo homework. Too much homework only antagonizes kids and, in most cases, it does not help their learning.

On the other hand, a small amount of homework is a good thing:

- It is a form of differentiation, as it allows kids to take different amounts of time to do the same assignment. (In a successful cooperative learning culture racing is discouraged, and kids work more or less at the same pace, with the faster students slowing down to help others when needed. This is not just altruistic, as doing this helps deepen their understanding.)
- It gets the message across that it takes work to learn anything substantial, and that while in class we work in groups, the ultimate goal is to understand the material well enough to deal with it on your own.
- It's a place to do the often necessary but often boring work of basic drill and review, thereby saving class time for more interesting and substantial engagement.
- Homework helps you sort out what you understand from what you are still struggling with, which will make it easier for you to get exactly the help you need when going over the homework with your group. (Peter Liljedahl suggests that homework problems should be called "Check Your Understanding".)

The message is: you need to know how to do this on your own, but you can get as much help as you need. You are not being rushed or pressured. Our goal as teachers is not to separate those who can from those who can't: it is to get to where everyone can.

Henri writes:

> A powerful argument in favor of homework came from the students themselves. In course evaluations, it was common for students to tell me that what helped them learn the most was going over the homework with their classmates. I usually allowed up to 15 minutes for that at the beginning of class. During that time, I walked around and recorded a 0 (did not do it), 2 (great job), or 1 (somewhere in between). Students helping each other is far more efficient than me explaining things to the class, because it allows different groups to focus on the parts of the assignment they each need to focus on. True, there's a risk that all the students in a given group are doing something incorrectly, but that doesn't happen very often. Besides, after a quick homework check, I'm walking around, ready to intervene if I see this.

(Henri taught in a block schedule with long periods. This approach may not work as well if the math period lasts 50 minutes or less.)

A teacher explanation at the board is sometimes needed, particularly if there is a similar question in more than one group. To make this more engaging to students who got it right in homework, it works best if the problem done at the board is similar to the one that stumped many students, but not exactly the same.

However, in general, it is not a good idea to go over homework problems: it is boring for students who did the work correctly at home, and often insufficient for students who really don't get it. It is also a waste of precious class time. Asking students to write solutions on the board does allow others to focus on the ones they need help with, but live conversation is more effective than silent copying of ill-understood techniques, and again, it is a waste of time for students who did the homework correctly. The bottom line is that the goal is not for every student to have the correct answer written down for every homework problem. The goal is for students (and their teacher!) to know what students can do and understand.

Still, you may be in a school where students do not or cannot do homework. Fair enough: homework is not a matter of principle. It is a component of a comprehensive approach, and the same goals can be accomplished in other ways. If a student cannot do homework because of their home situation, the school should offer study hall time so that "homework" can be done at school. If the school cannot or will not do that, and many students are unable to do homework, then some class time can be allotted for that: time where you work on your own, after plenty of time working with peers, and prior to being quizzed. Of course, giving up class time for this will reduce coverage, but it may be a price worth paying.

However, we have heard from more than one teacher that homework completion increased dramatically once homework was lagged. It turned out that the reason many students weren't doing the homework was because they didn't know how to do it. Lagging gave them time to learn the concepts, and made homework possible. So before abolishing homework, you should try both lagging it, and offering in-school time for it. More students will learn more math than with no homework at all.

## Review

We have mentioned review as a normal part of teaching math many times in this book. However, *approaching the same idea the same way is not effective*, no matter how many times you do it. It becomes boring and students tune out, which undermines the goal of the repetition. Instead of blaming the students for not benefiting from broken-record performances, it is more effective to expand one's teaching repertoire, and get to know multiple ways to teach important topics – look back at our arguments for and examples of learning tools and multiple representations in Chapters 5–7.

If your textbook does not include that sort of variety, you'll have to learn new things outside the textbook (including from other textbooks!) This cannot be rushed, as there's only so much time in the day and the week, but you can gradually expand your horizons by attending conferences and workshops, by internet searches, and especially by talking to colleagues. Learning new ways to teach the same content is a career-long project.

Review for the purpose of maintaining a set of automated skills is only a small part of the reasons for engaging in it. For some students, it's an opportunity to learn a concept they didn't get the first time; or make connections between that concept and other parts of math, or iron out misconceptions, or apply the concept to a specific context, or get to where they can explain it or generalize it (the meta-curricular goals of articulation and abstraction).

Done poorly, review can kill. There is nothing more deadening to a class than being told on day one of a course or unit that a week will be spent on a review packet. It is especially catastrophic to start Algebra 1 with a review of arithmetic. Students who have mastered the material find it insulting, or at any rate, boring. Students who have not find it demoralizing. We understand why that is a standard practice, but we believe it is counterproductive. Over time, it tells students "You don't need to remember anything – I'll make sure to remind you".

If not at the start, then when should review happen?

- If your students do homework, that's one place to put review. Use short assignments that hit the key ideas they will need.
- If only a few students need the review, find ways to do it outside of class in some sort of support system offered by you and your colleagues.
- If most students need the review, it will be obvious to all and it will not be resented. It should happen as soon as the need arises.

**More on Review**

Review is necessary, but how does one make it both palatable and effective? Lagging homework is one way: any given topic is reviewed in homework sometime after it is introduced. Implementing such a policy is not difficult, and it has substantial benefits. But there are several more ways to build in review:

- Separating related topics.
- Pursuing two units at any one time.
- Lagging assessments.

- Cumulative assessments.
- Quiz and test corrections.

In combination with each other and with lagging homework, these strategies can dramatically extend student exposure to important ideas, without interfering with constant forward motion. We will discuss them in the next two chapters.

## Discussion Questions

1. What is your homework policy? Does homework "count" in determining student grades? Should it? How do students know if they did the homework right? Can lagging homework fit in your system? Are any other changes needed?
2. Review is more effective if it involves a different approach than you used when the topic was first introduced. Choose an important topic where you think some review is in order, and discuss some ways to come at it using a different representation, a different context, or a different tool.
3. Competition between forward motion and process is most evident when asking students to explain and justify. In what ways is this a trade-off? In what ways is it a false dichotomy?

## Note

1. See Henri's entertaining five-minute video on time pressure: mathed.page/teaching (scroll down).

# 10
# Planning

Lagging homework can be done no matter what your overall plan for a course or unit is, which is why we presented it first. Our other strategies for extending student exposure to important ideas are closely related to assessment, which we address in the next chapter, and big-picture planning, which is the subject of this chapter.

## Pruning

Students need to see important topics repeatedly. This is most effective if one is using different tools and representations. But there is no time for that if the syllabus is bursting at the seams with too many topics. Thus, it is important to prune the curriculum. (Even if you do not have the option of taking topics out, you might still prioritize certain items, and allot time accordingly.)

Almost certainly, many things can go. Any topic that is only there because it is a personal favorite should at least be considered for deletion. Any topic that is unconnected to things that come before or after can probably be skipped. Any topic that is only there out of habit should go. Any topic that only reaches a handful of students should be taken out as it is wasted time for the others. (Do those super-hard topics in the math club, or move them up the grades to where they have a chance.)

Most controversially, you should consider spending less time on techniques and procedures that have been rendered obsolete by the availability of technology:

- Less time on paper-pencil multi-digit multiplication and division.
- Less time on complicated factoring problems and techniques.
- Less time on complicated equation-solving.
- Less time on complicated manipulation of radicals.

The reality is that for most learners many of the more complicated techniques, and the hours of practice needed to master them, do not shed any light on the underlying concepts. Quite the opposite: in some cases, they have been designed precisely to get accurate results without having to think or understand.

You may feel we are reckless in making these suggestions. We are not. The concepts behind these manipulations remain essential!

- Multiplication and division will of course always be important, and there are many interesting things you can do to help students make sense of them, for example exploring arrays with Base ten blocks, estimation, and mental arithmetic.
- The concept of factoring polynomials will always be important, and a student who cannot factor anything does not understand the distributive law. That is pretty much catastrophic from the point of view of developing any sort of symbol sense. The rectangle model can help, perhaps using Lab Gear or some other algebra manipulatives.
- Building connections between equation-solving, graphs, and tables is a good way to develop sense-making. Also: solving simple equations mentally is for some reason not a standard activity. It should be!
- An interesting way to talk about radicals is geometric.[1] Simple manipulations of radicals remain important for communication and reasoning. For example, it is useful to understand that $\frac{1}{\sqrt{2}} = \frac{\sqrt{2}}{2}$, if only to confirm two answers to a geometry problem are equivalent.

These concepts will always be important, so you should of course still teach them. But you will reach more students with deeper understanding if you prune your curriculum. One step at a time, and in consultation with your colleagues, cut back on complicated manipulations, use technological tools appropriately, and do more interesting, engaging work.

Once you have pruned your curriculum, you need to think about sequencing.

## A Framework for Planning

Figure 10.1 represents traditional pedagogy. The teacher explains the concept clearly, and students get more comfortable with the concept by practicing specific skills. Solid skills and clarity on the concept make it possible for students to build on that foundation and apply the skills and the concept to new problems.

**Figure 10.1** Traditional pedagogy.

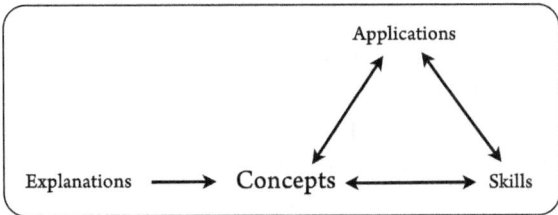

As you probably realize by now, we do not think it is that simple. Figure 10.2 represents one way to organize many of the ideas in this book into a useful framework:

**Figure 10.2** A better model.

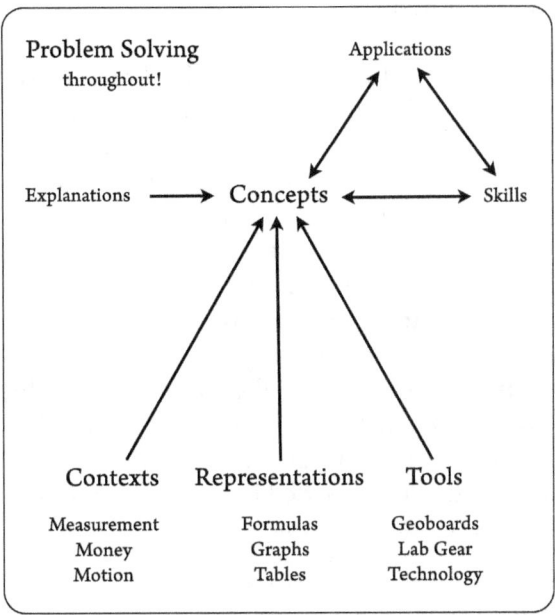

The idea is not merely that doing more things is better than doing fewer things. As we see it, concepts are best developed by spending a lot of time at the bottom of the diagram. The interaction of contexts, representations, and tools provides a foundation on which concepts can develop. It makes it possible for students to understand teacher explanations. This in turn makes the applications of concepts possible, and the practice of skills meaningful. As the arrows indicate, all six ingredients contribute to the development of the concepts – all the more so if problem solving is infused throughout.

Planning ♦ 179

Multiple representations and tools, two pillars of our edifice, are related. (In fact, one could argue that representations are tools, and vice versa.) We discussed both at length in Chapters 5, 6, and 7, but we should probably say more about contexts. The most fruitful contexts are concrete and beg for mathematical interpretation. The examples given in Figure 10.2 (measurement, money, motion) are particularly fertile for learning, but of course, they are not the only options.

Well-chosen contexts ground the math in students' experience. Using them is helpful in several ways:

- It reveals that math is about tangible reality, which helps to earn student buy-in.
- It helps solve the word problem conundrum: starting from contexts is a lot more effective than starting from rules and formulas.
- Most importantly, it increases student understanding and engagement with the ideas by allowing them to use their intuition in a concrete environment.

Note that in our view, contexts need not be "real world": they are not required to be strictly utilitarian or realistic. Area, while obviously connected to many real-world situations, works well as a context for mathematical exploration or reasoning, for example in a model of the distributive law, or in proofs of the Pythagorean theorem. Or take the story of the inventor of chess requesting payment in the form of one grain of rice on the first square of the chessboard, two on the next, then four, eight, and so on. This is a valid context for an introduction to exponential growth or geometric series, but there are many ways that it is not "real world".

In any case, when we emphasize the importance of the bottom half of the diagram, we do not mean to imply that the top should be thrown away. Quite the opposite: we propose all this foundational work to make sure that teacher explanation and skills practice have a chance at being effective!

**Forward Design**

This framework can be used to help evaluate a unit plan, a textbook, or the math program at a school. Is problem solving at the core? Are there enough activities involving contexts, representations, and tools to lay the groundwork for student understanding? The framework can also be used in *forward design*, an approach to unit planning proposed by Carlos Cabana, a math teacher in Oakland. When designing a unit, he suggests that a teacher can start by asking these questions, preferably in conversation with colleagues:

- What are the big ideas? What are all the different representations of those ideas? How can students consider it geometrically,

graphically, numerically, algebraically, and verbally? If you can only come up with specific micro-skills, think some more: what underlying concepts connect these skills?
- What ideas should students be able to justify and generalize? (Generalizations without accompanying justifications can lead to memorized and misremembered formulas – exponent rules being a prime example.) How should they be asked to reverse processes?
- What manipulatives or other tools are available to provide a way for students to engage in thinking and exploring? The right tool can make it possible to formulate a question that all students can engage in; it can support reflection and discussion; and it can add variety to your course. For example, basing algebraic symbol manipulation on the Lab Gear lets us ask algebraic questions about geometry and geometric questions about algebra.
- What contexts are there that can yield useful problems and significant understandings? For example, earning money helps students make sense of slope as rate: a slope triangle on a graph with a rise of 20 and a run of 5 can be interpreted as earning $20 every 5 weeks, or $4 each week.
- What curricular resources can complement or replace the textbook? Look on your shelves, search the Web, and ask your colleagues. This step is crucial, as you most likely do not have time to create everything from scratch, and moreover, freshly-minted activities usually require some classroom testing and tweaking. However, it is not enough to find a cool activity: it is important to think about how it will connect with everything else. Moreover, just looking at contemporary materials can prevent you from casting a wide net that includes older but still powerful lessons and even units.
- How will the students be working at different stages? Individually? In pairs? In groups? In whole-class discussions? Different modes are appropriate to different activities, and doing everything in a single one of those is a costly mistake if you aim to avoid boredom and want to reach the full range of students. What are the opportunities to delegate authority to students, so that we have the room to be surprised and delighted by what they do?

The planning of individual lessons flows from these guidelines.

**Brainstorming**

A possible way to implement forward design is to start by brainstorming about the questions in the first three bullets above, and to map out connections. Figure 10.3 shows an example of a "mind map" built around the context of area, in a possible algebra course.

**Figure 10.3** A mind map for area.

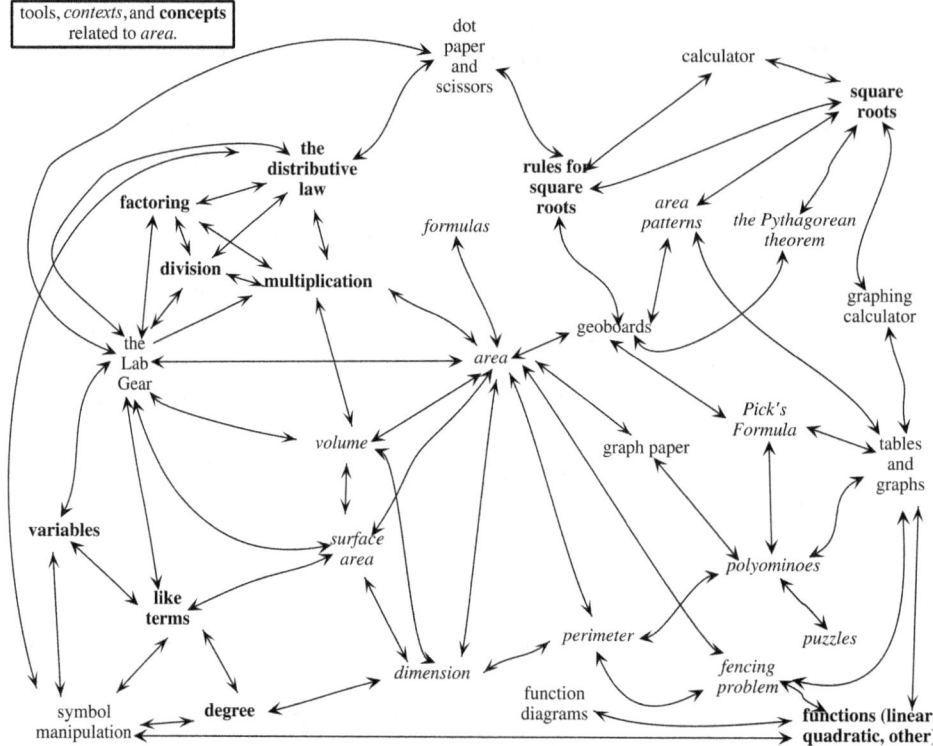

Admittedly, this is going overboard, and we are not suggesting this should be done every day, every week, or even for every unit. Brainstorming is something you might do at a meeting with colleagues, perhaps in the case of a brand-new course. We share this map not so much so you will study it closely, but to make the point that math curriculum is not a simple, linear sequence of topics, as could be suggested by looking at a syllabus or table of contents. It is an intricate web of tools, contexts, and concepts, and it is a good idea to spend some time thinking about how they connect before jumping into unit design and lesson planning.

Of course, as Bill McCallum pointed out in a blog post,[2] math is not linear, but instructional time is linear: one needs to eventually make some decisions about the sequencing of topics and lessons.

In any case, the beauty of this multidimensional approach is that it allows needed repetition without boredom: each time the subject is approached, it feels different because of a different context, a different tool, or a different mathematical connection. This gives weaker students many opportunities to get on board, and it keeps things interesting for stronger students. This works much better than doing the same thing again if it did not work for everyone the first time. Let's face it: doing the same thing will usually yield the same results.

## Mapping Out a Course

Effective sequencing, like everything else in teaching, requires paying attention to what happens with actual students. Whatever sequencing you end up with, consider it provisional, and don't lose sight of the reality that if it does not work, you can and should change it. There is no foolproof sequence for teaching math. Here are some hopefully helpful guidelines.

First decide what the **main topics** would be. The main topics, not all the topics you can think of. Even if you're forced to work with an overstuffed list of standards, the first step is to prioritize.

Organize the main topics into **units**. In general, something that doesn't deserve a unit or fit in one probably shouldn't make it into the program. Also save a stretch at the end for review, or just as cushion, if things take longer than expected.

- *Can the topic fit in a single unit?* Some topics are important but difficult for some students. It's a good idea to spread those out over more than one unit, and sometimes more than one course. One example of that is linear, quadratic, and exponential functions, which can be approached in different ways at different levels, from Algebra 1 to Precalculus. Another example is trigonometry, which can be distributed among Geometry, Algebra 2, and Precalculus.
- *When in the school year?* Difficult and important topics should not be introduced in May! By then both students and teacher are tired, and there is little chance of success. Teach those topics as early as possible. There may also be some traditionally late topics that can be useful early on. For example, in a geometry class, an exploration of inscribed angles has only very few prerequisites, and can be used to practice angle basics in an interesting context.
- *Tradition and logic are not your best guides*. Many textbooks are sequenced the way they are because that is the way it's been done before. That is not a sufficient argument for that sequencing. Also, do not expect mathematical logic to provide the best pedagogical sequence. For example, it is perfectly legitimate to address the Pythagorean theorem early on, based on an informal discovery of the formula, and then revisit it later in the course with a more rigorous proof.

How about lesson sequencing within a unit?

- *Concepts first, vocabulary and notation later*. For example, the tangent ratio can be introduced with the help of slope, without having to mention trigonometry or the calculator. Once students can use

the concept to solve problems, you can name it, present the usual notation, and reveal that actually, there is a key on the calculator for it.
- *Do not necessarily start with definitions.* Most students find it difficult to understand a definition for something they have no experience with. It is more effective to start with activities leading to concepts, and introduce formal definitions when your students have some sense of what you're talking about.
- *Concrete or abstract?* Math is all about abstraction, but understanding is usually rooted in the concrete, so it is usually a good idea to start there. This can mean many things:
  - Discrete first, continuous later; natural numbers first, real numbers later; numerical examples first, generalization later. For example, work on the geoboard is strictly about specific examples based on the available pegs. But it lays the groundwork for a generalization using variables which would otherwise be impenetrable to many students. Likewise, almost any new idea is more accessible if you start with whole number examples – as in $\sqrt{4} \cdot \sqrt{9} = \sqrt{36}$.
  - Kinesthetics and manipulatives do not accomplish miracles, but they can improve classroom discourse and provide meaningful and memorable reference points. In particular, algebra manipulatives can provide both access and depth to an essentially abstract subject, by way of a visual/geometric interpretation.
  - Tables and graphs can help provide a concrete foundation for the study of functions. This is sometimes described as modeling: you start with a concrete situation, use tables and graphs to think about it, and generalize with equations.
- *From easy to hard?* Well, that is certainly implied in the previous bullet point. However, we will now challenge that assumption. (As it turns out, sequencing curriculum doesn't lend itself to simple choices.) It is a good idea to start with somewhat challenging material, then ease up, and keep alternating between hard and easy. Starting too easy can give the wrong impression, that the unit will not require work. In fact, most of the above guidelines are best implemented as a back-and-forth motion: for example, after introducing vocabulary and notation, one needs to re-introduce the concepts. Likewise for most of these guidelines.
- *Start with an anchor.* If you can, start the unit with an interesting problem or activity (the anchor). It should be motivating and memorable, and it need not be easy. A good anchor brings together key content with good practices, and generates curiosity and engagement. It is something you can refer to later on, to remind students of the basics of the unit.

*Geometry: A Guided Inquiry* (a book we mentioned in Chapters 2 and 4) makes good use of anchor problems. Chapter 1 starts with the "burning tent" problem: a camper who happens to be carrying an empty pail near a straight river needs to run to the river, fill the pail with water, run to the tent, and put out the fire. What is the shortest path to accomplish this? Chapter 2 starts with the question: which polygons tile the plane?

Henri started a unit on exponential functions with an experiment.[3] Each group has 40 ten-sided dice in a shoebox. They shake the box, remove all the 0's, and count the remaining dice. They keep doing this until they are out of dice. The groups' numbers are averaged, the results graphed, and the class discusses how to write a formula for what happened. Students are stunned at how closely the formula's graph matches the class data. The payoff: later on, reminding students of the experiment is a powerful way to help them see the role of $a$, $b$, and $x$ in the equation $y = ab^x$.

## Spiraling

Repeated and extended exposure to the most important topics should be a key ingredient in your plan. However, spiraling can be overdone, and result in an atomized curriculum consisting of chunks that are so small that students don't get enough depth on each topic. Also, in some spiraled curricula the topics addressed by each problem or each lesson are not always explicit. This makes such programs more difficult to use. On the other hand, it takes a lot of work to create a spiral approach to a traditional, not-spiraled textbook.

Most of us teach related topics consecutively, hoping that if the previous topic is still fresh in the student's mind, it will make the next one easier to grasp. Examples are teaching multiplication by a two-digit number right after multiplication of any number by a one-digit number; factoring right after teaching the distributive rule; or sine, cosine, and tangent in one fell swoop.

Well, most of us are wrong about this. It is far more effective to *separate related topics*. If you do, when introducing the second topic, there is a built-in review of the first one. And thus, you have created forward motion (moving on to something else after introducing the first topic), and extended exposure (revisiting the first topic when working on the second.) Another advantage of this approach is that it communicates to the student that they are learning for the long run: any important topic may and will come back. All the examples above can be done in this manner: many-digit by one-digit multiplication, then later a more general multiplication strategy; comparing fractions, then sometime later, fraction addition; the tangent ratio, then sometime later, sine and cosine. The gap can be anything from a few weeks to a whole semester.

Note that it does not take more time to lag homework or separate related topics. It is merely a rearrangement of already-scheduled time, a rearrangement that supports both forward motion and review. In fact, it is an easy-to-implement form of spiraling. The idea behind spiraling is that students benefit from seeing important ideas more than once. That is absolutely true, but a poor implementation of this concept can backfire. In homework, for example, assigning one exercise each from several disconnected topics is not a bad way to check on student recollection of the topics, but it makes it impossible to assign a coherent set on exercises that build on each other.

The main problem with over-spiraling is the above-described impact on learning. But do not underestimate its impact on the teacher. For example, some spiraling advocates suggest homework schemes such as "half the exercises on today's material, one quarter on last week, one quarter on basics". Frankly, it is not fair to make such demands on already-overworked teachers. Complicated schemes along these lines take too much time and energy to implement, and must be re-invented every time one makes a change in textbook or sequencing. Those sorts of systems are likely to be abandoned after a while, except by teachers who do not value sleep.

The spiraling strategies we propose, in contrast, are easy to implement, preserve the integrity of instructional units, and offer the "eternal review" advantages of spiraling.

## Pursuing Two Units Simultaneously

One more strategy to extend exposure: *focus on two units at any one time*. For example, here is the outline of trimester 2 in a "Math 2" class:

| *Weeks 1–3* | *Weeks 4–6* | *Weeks 7–9* | *Weeks 10–12* |
|---|---|---|---|
| Circles | Quadrilaterals | Construction | Review |
| $n$th Power Variation | Systems of Equations | Trigonometry Intro | Review |

(Our point is not to recommend this exact sequence, which depends on many department-specific assumptions, but to use it as an example of what is possible.)

Here are some of the advantages of that approach:

- Any one day or week is more varied, which is helpful in keeping students interested and alert. To maximize this advantage, it helps to match topics that are as unlike as possible.
- It takes roughly twice as many days to complete a unit. This is good for students who need that extra time.

- This takes nothing from students who pick up ideas quickly. In fact, they appreciate the variety.
- It makes it easier to balance challenging and accessible work: if you hit a difficult patch in one topic, you can ease up on the other one. More generally, you gain a lot of flexibility in your lesson planning.
- If your work on one topic hits a snag, you can emphasize the other topic while figuring out what to do.
- Perhaps most importantly, it carries a message to the students: you still need to know this when we're working on something else.

If the school schedule includes long periods, it is possible to hit both units in one period, for example by introducing new ideas on one topic during the longer part of the period, and applying already-introduced ideas on the other topic in the remaining time. (Homework is typically on one or the other topic, not both.)

Can this approach, or a version of it, be used in traditional 50-minute classes? We guess that the answer is yes, but we have not tried it. It might involve, for example, focusing on each topic on alternate days, while the homework is on the other topic.

But, you ask, is this not confusing to students? Don't they prefer focusing on one single topic? If they do, that is only because that is what they're used to. In teaching, the biggest obstacles to making change are the cultural ones: the expectations of students, colleagues, parents, administrators, and of course one's own deep-seated habits. If working on two units, separating related units, and lagging homework are department-wide policies, students get used to it and do not question any of them.

## Learning from Experience

As should be clear as we approach the end of this book, teaching is a complex endeavor. The plans you make do not necessarily turn out to be right. The ideas you and your colleagues come up with may need adjusting, or may even fail completely in the real-life context of your district, school, or department. This is true whether you came up with those ideas in the summer, at the beginning of a term or unit, or the night before a lesson. The same, of course, is true of the prepackaged plans that are embedded in textbooks and curricula – probably even more so.

This is why the administrators who insist on seeing lesson plans in advance are barking up the wrong tree. Given all the variables involved, it is not possible to assess the quality of a lesson plan, and there is no guarantee that the implementation will match the plan.

We encourage you to keep detailed notes at the end of each day (or perhaps more realistically at the end of each week) about how things went.

These will be invaluable the next time you make plans to teach these lessons, units, or courses. After repeated adjustments, a course will approach a fairly stable state. Alas, by then, things may have changed: a different student population, a different set of colleagues, new developments in the software, a new edition of the textbook – any of those things may send you back to the drawing board! That is the nature of the job: as you approach your goals, they seem to recede in the distance. As a teacher, you must be a lifelong learner.

## Discussion Questions

1. If you will soon be starting a new chapter or a new unit, how can you make sure you will launch it with an interesting anchor rather than a boring review of old material? This will likely require cleverly scheduling any necessary review prior to the launch, or soon after. Discuss your plan!
2. For an important concept, discuss implementation of the forward design scheme we suggest above.
3. One of the main challenges all math teachers face is the fact that students pick up concepts at different rates. We have addressed this in various ways throughout this book. Look over previous chapters to find ways we address classroom heterogeneity. Discuss some of them.
4. Choose a course. What topics or techniques could be eliminated or shrunk to make room for more effective teaching of the most important concepts?

## Notes

1. See Henri's *Geometry Labs*, 8.4–9.4 (mathed.page/geometry-labs).
2. mathematicalmusings.org/2017/11/01/math-is-not-linear-but-time-is
3. mathed.page/alg-2/exponential

# 11
# Assessment

A balancing act faces all teachers: how much time should you spend grading? How much time should you spend planning? In the first year or two (or three) of one's career it is difficult to think clearly about this. The main thing is to survive the day, the week, and the semester. But maintaining an unrealistic workload is not sustainable in the long run: it will push you out of the profession sooner rather than later. If you want to stay in the classroom over the long run, you will need to balance the different parts of the job and accept that perfection is not going to happen.

*When grading, you are working for one student. When planning, you are working for the whole class.* Keep that in mind when you are budgeting your time. Don't grade more than you need to. Can students correct their own or each other's work sometimes? When grading, do you really need to write a lot? Do your students read what you write? Do they heed what you write?

That said, it is politically impossible to avoid assessment, as it is a major preoccupation of students, parents, and administrators. Since early in the 21st century, a draconian system of high-stakes tests has dominated the educational landscape in US public schools. This, to say the least, has not made teachers more effective – quite the opposite. Moreover, teachers have little say on how to adjust to this state of affairs. All we can suggest here is that you can try to minimize the damage by using some of the techniques we shared in this book in order to make whatever test preparation you have to do actually serve some educational purpose. In any case, this is not what this chapter is about: we focus on what you can control.

It is impossible to teach without assessing student understanding. Here are some legitimate, even essential uses of assessment: fine-tuning the course; helping both teacher and student know what the student understands and can do; and offering learning opportunities. All these are best served by decreasing the stakes. Lower stress translates into more accurate assessment. Ungraded formative assessments can serve the most important goals of assessment and should play a bigger role than they do in many classes.

We will share some suggestions on how to make grades and the whole assessment enterprise more accurate and less toxic. This is a charged topic, which is very dependent on department, school, and community culture. As with every other topic in this book, we can only address it based on our experience. In particular, we do not know a lot about standards-based grading, which is recommended by many people we respect, but which we've also seen implemented in ways that are counterproductive. As you read this chapter, we ask you to focus on the underlying ideas rather than the specific details.

## The Future of Grades

As of this writing, 370 schools throughout the country have committed to rethinking the high school transcript from scratch, leaving grades and GPAs behind. They organized themselves in a consortium, and are hoping to show that a different approach is possible, and that their work will eventually inspire change throughout secondary education. They are motivated in part by concerns and values similar to the ones we discuss in this book. They also hope it will help reduce the high level of stress experienced by high school students, and be a better tool for college admissions offices. You can read more about this on their website.[1] We wish them well, but of course, for the immediate future, grades remain central in most schools.

## Tests and Quizzes

First of all, we recommend that tests should weigh less in a student's grade. Tests are one of the most efficient ways of determining grades and communicating them to others. Unfortunately, you get what you pay for. Teachers who talk to their students usually know a lot more about their capabilities. Whatever subjectiveness arises in the use of this information is nothing compared to the loss incurred by ignoring it.

Still, much can be said in defense of traditional tests and quizzes: they provide a lot of information, they are easy to grade, and they are expected by all constituencies. Given that they are here to stay, how can we make them more effective? Here are some suggestions:

- Within reason, *give students as much time as they want*. If a student is not fast or does not do well under time pressure, so what? It does not mean they don't understand the material. Racing belongs in PE, not in the classroom. As far as we can tell, timed tests are strictly the result of bureaucratic efficiency. Students get in and out on schedule and the quantity of work for both student and teacher is limited to what the typical student can do in 45, 55, or 90 minutes. It is hard to think of an example in which the ability to solve a battery of problems in 90 minutes is more valuable than the ability to solve the same problems in, say, a day. Get rid of timed tests!

- *Do not over-penalize students for small computational errors*. Yes, precision matters, but if those errors could be eliminated by the use of technology such as calculators and computer algebra systems, we are less concerned about them. Prioritize evidence of understanding, not nit-picking accuracy. (Yes, sometimes computational errors reveal a lack of understanding, and of course that is not what we're talking about here. Paying attention to those mistakes over time may yield useful information along these lines.)
- Getting the right answer matters, which is why one might give less than full credit in the case of small errors. But if accuracy really matters, *allow any and all technology during most tests*. (Yes, there is an important role for no-technology tests, but they should not be the default.) Such a policy has salutary consequences on the selection of problems: deemphasizing procedures, emphasizing understanding.
- *Lag quizzes and tests*: give new topics a chance to settle into the student's consciousness before testing them.
- Periodically (or routinely), *administer cumulative tests*, which include topics from earlier in the course. This is consistent with the "eternal review" concept we promoted in Chapter 9. It communicates to students that learning is for the long haul. Especially in combination with test corrections, it helps to reduce the stakes: students get more than one chance to show their understanding of a given topic, and midterms and finals are not as exceptional and intimidating.
- *Include "bonus" questions*, which are important to challenge your strongest students, and which can be used to deepen or extend understanding. Those can be required all students in the test corrections. This gives the message that getting 100% is not easily achievable, and keeps everyone from getting complacent. It also helps to communicate that a test is a learning opportunity. There will be pushback on this ("This is not fair!") but in fact, what would not be fair would be to limit tests to questions everyone can answer, as it would lower course expectations. Working hard on those items as part of the test corrections makes everything else more accessible. Of course, such problems should not carry much weight in scoring the test: one should be able to get a top grade without solving them.

- Use *participation quizzes*, during which the teacher watches the class and visibly takes notes on students' desirable behaviors where all students can see them, for example by projecting their computer screen. This is a surprisingly effective technique to promote the most productive ways to function in a math class. As students see them on the screen, they emulate those behaviors in the hope of being mentioned and improving their grade. (It is difficult to not get an A on a participation quiz.) For the purpose of this activity, they should be asked to work a reasonably accessible assignment. Here are notes of the kind we are thinking about: "Jared took out his materials right away and started working. Shanda is helping Gerry with a challenging problem. Janine is moving closer to the group so she can hear those explanations". And so on. Students are being assessed on work habits, not math understanding, but one leads to the other.[2]

## Retakes Versus Test Corrections Versus Neither

To the above list, we add:

- *Make test corrections "count"*. This lowers the stakes in a good way. Everyone knows that doing well the first time is better, but if learning is the goal, what difference does it make if the learning occurs a week later? Students can get help from anyone, but all writing must be their own. This assumes a standard of explanation that is higher than on the test itself.

This suggestion falls under a common topic of discussion among math teachers: the question of "retakes". Under what conditions should students be allowed to have another chance at taking a test? How does the retake affect the grade? This is an important conversation. Different opinions reflect different values, different attitudes toward assessment, and different understandings of how learning happens. We will present this as a discussion with imaginary colleagues, whose contributions are in bold type.

*– No retakes! The test is an accurate assessment of the student.*

Doing well on a test does usually reveal a mastery of what was tested. However, doing poorly is not as reliable an indicator of understanding, as the problem may be that the student needed more time, was not feeling well, or made so-called "careless" mistakes in spite of having a decent understanding of the material. Given all this, we cannot accept that argument.

*— If they did poorly, well, they should have prepared better. They'll never learn responsibility if they keep getting second chances.*

Students often say they don't know how to prepare for math tests. They don't have this issue in other disciplines. The reason is that doing well on a worthwhile math test requires understanding the underlying concepts. Most students cannot improve their understanding by "studying" – unless the test prioritizes memory over understanding.

Even if they have the maturity to struggle for understanding while preparing for the test, it may well be that they need help in order to do that successfully.

*— They need to be trained to the harsh reality that colleges do not offer such opportunities.*

Bad pedagogy in college is not a valid justification for bad pedagogy in high school.

*— Getting a bad grade will help them take the next test seriously.*

More likely, it will convince them they're not good at math, especially if they see that their teacher considers the test to be an accurate assessment of their ability.

*— In any case, I don't have time to create, schedule, proctor, and grade retakes!*

That is a valid argument, to which we return below.

Some teachers allow retakes under certain conditions:

*— I allow retakes if the score was below a certain threshold*
*— I allow this many retakes per grading period*
*— I allow retakes if the student has shown they are serious (e.g. done their homework, etc.)*

Such policies are more complicated, and less extreme than a simple "no retakes" system, but they are justified with the same arguments. The idea is still that the number of retakes should be reduced, though not completely eliminated. This sort of thinking is based on some assumptions, which are not stated explicitly, and don't need to be because they are nearly unquestioned. Here they are: (1) students should have mastered the material by the time the test comes around; and (2) they should be able to show this mastery under time pressure.

But everyone knows that in reality, students learn new math concepts at different rates. And while some adrenaline-fueled students thrive under time pressure, others freeze and get anxious. Those character traits may or may not be related to the student's understanding of the material. When students tell us they need more time (more days before being quizzed, or more minutes for the quiz), they are often telling the truth. Is understanding achieved a few days later less valuable? No! Does needing a few more minutes to figure something out reveal an inferior understanding? Why should it matter that Student A could finish the test in 50 minutes, and Student B needed 65 minutes?

We need a system that:

- prioritizes student learning (not concerns about grades);
- does not involve extreme time pressure;
- respects teachers' time.

*– So, what do you suggest?*

One way to achieve all three goals is test corrections, done as homework. At Henri's school, teachers called this a "recycle" of the test. Here is a possible recycle policy:

- Expect a high standard of explanation, higher than what can be accomplished on a timed test. Ask students to explain it well enough to convince you they understand, and especially to convince themselves they understand. Do not ask them to dwell on their mistakes, as we don't think it's helpful, and it may even be counterproductive.
- They have a week to do it. They can get help from a teacher, or each other, or a tutor, but everything must be in their own words, and all helpers must be listed.
- All mistakes should be corrected by all students, whether or not this will affect the grade. (The main purpose of the recycle is learning, not grades chicanery.)
- Extra credit/bonus questions should be recycled.
- A perfect recycle gets the student's score halfway to 100%. (Or, some other version of this that is consistent with the department's grading policy.)

*– How does this respect teachers' time?*

Well, it does not require creating another version of the test. It does not require scheduling and proctoring the retake. And the grading is extremely fast: remember you're only going over problems you have graded before, and moreover, only a small fraction of them, since most students are only recycling a small number of problems.

*– But this does not tell you how the student would do on a timed retake!*

So what? In many disciplines, students are assessed without being subjected to time pressure. Why not do that in math? (Besides, we have already had a timed test, the one that is being recycled!)

*– But if they are allowed to get help, this does not assess what they can do on their own!*

Seeking help when you need it is a good thing. A student who works hard on the recycles and turns in high-quality explanations in their own words is sure to learn a lot. Isn't that our goal?

## Other Assessments

An extremely effective approach to assessment is one-to-one work with a student. Frustrated by the inaccurate results yielded by a standard placement test, Henri changed his department's policy. Under the new system, the student took the test with a math teacher sitting by their side. The teacher could answer questions, give hints, and generally probe for understanding. This approach was not infallible, but it was more effective than just administering the test the usual way. Because it was labor-intensive, it was only used on students whose placement could not be determined by transcript and standardized test results alone.

Robin had the same insight. He writes:

> In a calculus class with 32 students, I decided to give a 15-minute oral midterm instead of a 90-minute written midterm. It took 8 hours of my time, but then so would writing a test, giving it, and grading it. Either way, an entire (long) day was shot. This taught me a huge amount about how the students thought and gave me a measure of certainty when I assigned grades at the end, that everyone was getting the grade they deserved. Students neither loved it nor hated it, but it worked well for them because of how it informed me.

In addition to better tests and quizzes, it is important to have significant at-home assignments, including especially the test corrections mentioned above. In schools where homework is not a realistic expectation, such assignments can be completed in study halls or in special class sessions.

Here are some possibilities:

- *Reports*. Ask students to summarize a unit in their own words and with illustrations. Keep those to a reasonable length: one or two pages, or a poster.
- *Projects*. For example, write a (very) short science-fiction story involving exponential growth, with an appendix explaining the underlying calculations. Or use Desmos to create a picture using graphs. Or use GeoGebra to construct an Archimedean solid.
- *Take-home tests ("problem sets")*. Those can and should be more difficult, and require more time, than in-class tests.

Students can get help but they must write everything in their own words. If they get help, they should list who helped them, and who they helped. But if help is available, how can one use these to help determine a student's grade? If sorting students was more important than teaching them, that would be a valid point, but on balance take-home tests do help student learning. A somewhat less differentiated set of grades is a price we're willing to pay. In fact, that very attitude helps to shift school culture, by emphasizing and communicating that learning is what is most important.

There is an equity argument against at-home assessments: students with access to tutors or math-savvy parents have an advantage. Still, the main thing is to relieve the time pressure, to get a better-rounded sense of the students' understanding, and to offer them another avenue to strengthen it. Each teacher needs to weigh these pros and cons and decide how much weight to give at-home assignments.

Those assignments often reveal that some of the students who do exceedingly well in a timed classroom test do poorly when the assignment requires a more thoughtful approach. This is important information for the teacher, and moreover, as long as we're trying to be fair, it levels the playing field somewhat.

In practice, test corrections are quick to grade, but reports and projects may not be. This should be taken into account when scheduling them. It may be that one per grading period suffices.

## Yet More Options

The above are approaches to assessment we are familiar with. However, there are other options. We will not say a lot about them, as we are not experts, but here are a few ideas[3]:

- *Group tests*, with the score determined by a random drawing among each group's papers. Those can generate productive collaboration and much learning.

- Observing and evaluating students' *class work*, as in the participation quiz, but routinely and silently.
- *Notebook checks*, which give you a different window on student understanding and work habits.
- *Holistic scoring* of student written work, which is a lot faster than using rubrics. Or, if you worry that such scoring is overly subjective, use a Yes/No rubric for quick assessment of take-home assignments. Some version of this is vastly more efficient than a complicated multilevel rubric.[4]
- *Portfolios*: a student-compiled folder containing the student's best work.
- *Student self-evaluations* in journals or other formats can help round out the picture, and provide the basis of a productive conversation. These work even better if you have asked students to set some personal goals at the start of the grading period.

## Assessment and Extending Exposure

Depending on school and department culture and values, not all of these ideas will apply. Perhaps totally different strategies are in order at any particular school, but no matter where, the general principles are: *reduce the stakes, vary the assessments, and prioritize student learning* over concerns about grading and ranking.

Also note that many of our assessment suggestions are consistent with our call for *differentiation by time* and *extending exposure*. This approach, in combination with the use of tools and multiple representations, makes it possible to have a "preview/view/review" cycle for the most important concepts. If you put it all together, a student will come across an important concept in class, then in homework, then on a quiz, then in quiz corrections. In some cases, the concept may come back in a cumulative test and its test corrections, and/or when starting a related topic. Some students will grasp the concept on Day 1, others will need some or all of these reviews. Exposure to the idea has been dramatically extended, while the course continued to move forward at an appropriate pace. Everyone wins.

One thing we did not address is standardized testing. We are confident that the pedagogy we have promoted in this book will lead to stronger understanding for more students. Stronger understanding will yield better test results. That said, you may need to find some compromise and reserve some time for test preparation. Do what you have to do, but do not lose sight of your most important goals!

## Discussion Questions

1. What is your policy on test retakes or corrections? Does it work to improve student understanding? Is it time to revamp it?
2. This chapter prioritizes the formative over the evaluative goals of assessment. Does your opinion about the ideal balance differ from ours and why?
3. Is there a way to incorporate some of the ideas from this book in work you do for standardized test preparation?
4. We recommend striving simultaneously for constant forward motion and eternal review. Summarize and discuss the techniques that enable these seemingly contradictory goals.
5. Is any version of at-home assessments an option at your school? Would occasional reports and/or projects be a worthwhile complement to quizzes and tests? What would be their impact on different students?

## Notes

1. mastery.org
2. We learned about this technique from Carlos Cabana, who attributes it to math educator Phil Tucher.
3. Anita Wah and Henri elaborated on some of these ideas in *Algebra: Themes, Tools, Concepts*, on pp. 552–555 of the Teachers' Edition (mathed.page/attc/tg).
4. See *Algebra: Themes, Tools, Concepts*, on p. 560 of the Teachers' Edition for an example (mathed.page/attc/tg).

# 12
# Making Change

Teaching for understanding, let's face it, is still countercultural in many math classrooms. Often, attempts at moving in this direction are aborted early, because they don't succeed immediately. Such is the power of culture. However, all involved are thrilled once there is success with the sorts of practices we outlined in this book. Liljedahl, the author of *Building Thinking Classrooms in Mathematics*, quotes a teacher[1]:

**Researcher:** How are you liking your classroom these days?
**Teacher:** I'm loving it. I feel like the students are completely different. I'm completely different. It's like I have a new job and its *way* better than my old one.

He found similar changes in attitude among students, who not only expressed enthusiasm about the changes, but also voted with their feet: absences went from an average of 3.2 per class down to 1.6; tardies went from 6.7 to 2.2.

But how does one get started? We end this book with some thoughts on how to move forward on some of the recommendations we made. Most of those were geared to teachers and departments – but all of us do our work in a broader context, one that involves tradition on the one hand, and new developments in the world of math education on the other. In this final chapter, we discuss avenues for improvement: feedback from students, collaboration with colleagues, and the never-ending stream of ideas from the broader math education community.

## Course and Teacher Evaluations

Our last chapter focused on how teachers assess students. We now turn around and focus on how to maximize the benefits of the feedback students offer in course evaluations.

Henri writes:

> Early in my high school teaching career, I was given a form that my students were supposed to fill out. It had a long list of questions, most of which struck me as unhelpful to me as a math teacher. All these years later, the only one I remember is "Did the teacher explain things clearly?" Even then, it was obvious to me that even the clearest explanations may not support student understanding. (There were no questions like "Did the teacher get you to engage intellectually?")

Over time he found that he got much better information by *not* trying to come up with a detailed list of criteria. Instead, his course evaluation form consisted of just three questions:

- What did you enjoy?
- What helped you learn?
- What are your suggestions for improvement?

The first two questions were intended to help students realize that well, those are different questions! Certainly, it's good for the teacher to know what they enjoy, but let's not lose track of the fact that the goal is to learn. Some students listed the same items in response to both questions, but most were quite clear on the difference. Typically, the first question elicited mentions of hands-on activities, while the second included things like "going over the homework with my group" and "working on test corrections".

As for the third question, students often offered responses that canceled each other out: go faster/go slower, more quizzes/fewer quizzes, and so on. But sometimes there was a preponderance of similar feedback, and then Henri knew it was worth attending to the issues raised.

This three-question form implies that students and teacher are on the same side. The positive tone helps to generate useful feedback. The wide-open nature of the form may result in surfacing issues the teacher had totally missed. It is most effective to do such a survey around the middle of the term. Waiting until the end of the course, when students have less at stake, and in any case when it's too late to implement any changes is problematic. Still, better late than never: the feedback will be useful in future iterations of the course.

That said, it is not realistic to rely exclusively or primarily on student input. It is best to complement that with classroom visits from a colleague – especially in your first few years as a teacher, and at any stage if you feel something is amiss. Such visits are especially useful if you ask for specific feedback: what is going on among the students during whole-class discussions? Do some groups seem to be off task? Or even: do I explain things clearly?

## Collaboration

In secondary and middle schools, the department is the main arena for teacher professional growth. Much can be accomplished there across different levels and courses, if the department is a professional learning community (PLC). In an ideal world, every math department is a PLC, but in reality, there are some obstacles to that idea:

- not all schools give teachers time to dedicate to professional learning;
- not all teachers are interested in professional growth;
- it is not clear what to do in a PLC, even if the first two obstacles are not a concern.

In-house professional learning is not in contradiction to the learning one can do online or by going to conferences. Learning from experts outside the school is a useful complement to what can be done onsite. However, an in-school PLC has enormous advantages:

- It is geared to a specific community of teachers, students, department, and school. Outside professional development resources may or may not be a good match.
- It can be coherent, while the catch-as-catch-can off-site PD opportunities are usually disparate and unrelated.
- It is a year-round opportunity for growth.

One of the biggest gains is in collaborative lesson planning. In elementary schools, there is more often coordination by grade level than there is a department structure organized by subject. When every teacher in a grade teaches math, there is a great opportunity to compare and discuss, plan lessons collaboratively, and occasionally see each other's classes. This is the setting in which lesson study was documented in *The Teaching Gap*.[2] In some elementary schools, teachers with an affinity for math semi-specialize by swapping: they will teach math in someone else's room, who in turn teaches English or social studies in their room. Regardless of the structure, unless the school is very small, there will always be more than one teacher teaching any grade's math classes, and therefore always opportunity for a PLC.

In middle school and high school, it works best if *different sections of a given course are taught by more than one person*. Those teachers meet regularly to map out the coming week – class work, homework, assessments, and so on. They discuss how those plans worked out in their classes. They think about the big picture of the course, and flag issues that will need to be addressed in future iterations. These collaboration meetings are not resented, quite the opposite! They feel useful and important.

This practice is more or less a prerequisite for lesson study. Under such a plan, no one gets to just teach one thing all day, so this involves a bit more preparation, but on the other hand, the preparation can be shared, everyone gradually gets to know the whole program, and teachers learn a lot from each other. Among other things, it allows the pairing of seasoned teachers with beginners, a much more effective approach to mentoring than occasional meetings about nothing in particular.

Over a few years, as teachers gradually cycle through the department offerings in different collaborative teams, ideas are shared, and everyone grows, especially if teachers take notes on what worked and did not work. Some summer collaboration on the following year's program can help put that information into shared documents, thereby institutionalizing what was learned. These practices gradually turn each course into a joint project of the whole department, a much preferable arrangement to individual teachers "owning" each course.

But what about turning the whole department into a PLC, in a way that does not require years to percolate? Here are some activities that can provide benefits when a department or grade level functions as a PLC for mathematics teaching.

- *Do math together.* For example, if one of you attends a Math Teachers' Circle,[3] bring the problems back to the department. Or find teacher-suitable problems online.[4]
- *Rehearse instructional routines.* This can mean anything from discussing to roleplaying.
- *Explore learning tools.* Take turns learning and teaching each other. Note that some tools may require multiple sessions of your PLC.
- *Read and discuss articles and blog posts.* There are actually dozens of math teacher blogs and websites, some of which are worth talking about. Or, if one of you is an NCTM member, you should be able to find many articles worth discussing in the NCTM journals.

An example of how some of these ideas might play out is provided by Kevin Rees, a math teacher in California.

> We have moved to a model with less frequent but longer meetings (1 hour to 2 hours depending on the schedule). During these meetings we like to start with a "warmup" (like we give the kids) – a sharing from a Math Circle, a conference, etc. where we all get to do math together. The beauty of this is that it also allows different department members to run parts of the meeting so that the meetings are a shared experience, not just the department chair running the whole thing.

What we have been doing this year has been powerful. We've been doing a deep study of just one chapter of a pedagogy book – *Becoming the Math Teacher You Wish You'd Had* by Tracy Zager.[5] I highly recommend this book and the website and materials that go along with it. We chose to work on Chapter 12: "Mathematicians Work Together and Alone" for the whole year – looking at how we give our students chances to collaborate, dig deeper, and also when to step back. It has been amazing – different teachers have been trying on different parts of the chapter (fully randomized seating every day debate structures, vertical surfaces, etc.) and then we report back to each other and hone the methods together.

If you are in a department of one, or if your colleagues will not work with you, all we can suggest is getting to know teachers at other schools, looking for collaborators on the Internet, and attending professional conferences and Math Teachers' Circles regularly.

## Research

Some might argue that math education research should be our guide. Research can help, but even valid research that is used to support the unrealistic claims at the core of various fads should not be generalized beyond reason. As one researcher puts it: "the issue is much, much less about the quality of the research and much more about how it is misused to create the fad". Fair enough. We agree on that, and we cited some such research in this book. So far, so good. But many math education research papers appear to contradict each other or to violate what teachers know from experience. Most teachers have neither the time nor inclination to analyze research papers in order to assess their validity.

For example, we heard that short periods are better than long periods for teaching math, because someone found a correlation between higher test scores and shorter periods. While we are sure some good teaching can happen in shorter periods, it's absurd to make a blanket claim that those are superior to longer periods.[6] If there is such a correlation, it could be due to the fact that it is schools that were less successful on standardized tests that felt a need to try block schedules. Or it could be that the tests measured something other than depth of understanding. Or it could be something else.

Here's another example. We are fans of the "growth mindset" concept, and in fact, have been making some of those points for years. Having some research to back this up is excellent. However the claim that being exposed to this idea in a slide show is sufficient to change a student's mindset is hard to believe, even if some short-term effects can be observed. What may have a lasting effect on student mindset is restructuring one's teaching to make clear in practice, not just in words, that students can get better at math. This would include such policies as extending exposure, lagging homework, de-emphasizing grades, valuing test corrections, making explicit every day that getting it wrong is often a necessary stage on the way to getting it right, and so on. In other words, a classroom culture that challenges the dominant culture. We would love to be able to point to research that supports those things.

We could look at any particular study and dissect it, but unless it is very unusual, it is likely that it shows that A was better than B for a certain population of schools, in a certain implementation, if you care about a certain measure, and measured it well. In fact, much empirical research in education operates under similar limitations. It's better than nothing, but you should be cautious about taking most of these findings too heavily into account when designing your own teaching.

The fact is that research is influenced and framed by the researcher's values. Teachers need to find math education researchers who share their values and use their results to refine their practice and to dialogue with administrators. For example, if equity is an important goal for you, find the researchers who share that concern. Their work is likely to be useful. If you prioritize understanding over memorization, or collaboration over competition, find the research that is about that. And so on. But to be honest, it will often be more fruitful and practical to get your ideas from fellow teachers.

Some years ago, at a conference, Henri saw a math education professor who was an expert on the learning of geometry. Having recently read one of his papers, Henri said to him: "Your research confirms my beliefs!" Without a moment's hesitation, the professor replied, with a big smile: "That's what it's for!" Was he joking? We don't think so. That is indeed what it is for.

## Fads Versus Eclecticism

During our four-plus decades in the classroom, we've seen many math edu-fads come and go: new math, individualization, manipulatives, problem solving, group work, constructivism, constructionism, portfolios, complex instruction, interdisciplinary-ism, backward design, coding, rubrics, problem-based instruction, technology, Khan Academy, standards-based grading, making, three acts, flipping, inquiry learning, notice-wonder, growth mindset, whiteboards… not to mention various generations of standards.

Fads will always be with us. We have learned to coexist with them. Many people want to help teachers, without being teachers themselves, so they contribute to that never-ending stream. We appreciate their efforts, even if we don't buy their often naive claims.

No one fad has all the truth, but most have some piece of it. Keep'em coming! We need to consider each one as it comes along. Instead of shutting our classroom door and continuing business as usual, we should keep it wide open. Without becoming a dogmatic across-the-board adopter of each pedagogical scheme, we need to learn what we can from it, and incorporate that bit into our repertoire. This is how we get the sort of flexibility that makes for good teaching. If we do that, our lessons will not fit a standard mold. Quite the opposite: they will depend on the myriad variables that make teaching such a complex endeavor.

In addition to the grand fads that periodically sweep through, we also have teacher-initiated ideas that spread through the grapevine, and sometimes find their way into curricula. Those are often more useful and less pretentious, as long as we avoid the temptation to see them as more than what they are. Teachers have little time for theory and are happy to collect interesting lessons from the Internet, from conferences, or from colleagues. This is true of all teachers: we don't have the luxury of constraining ourselves to a single theory, because our work does not allow it. We get ideas from many sources, and evaluate them by using them. The best of those ideas end up in our repertoire. Different teaching challenges lead to different pedagogical responses. We have no choice but to be eclectic.

**There Is No One Way**

According to Merriam-Webster:

**Eclectic:** selecting what appears to be best in various doctrines, methods, or styles

That pretty much describes our stance as educators.

It doesn't take long for a conversation between teachers to include something sarcastic about the fad du jour. By being sarcastic, we put up an umbrella and try to protect our sanity from the ideas raining on us from administrators, academics, and yes, even colleagues. We will go further, and boldly say to the proponents of the current pedagogical panacea: we're sorry, but whatever "evidence-based" universal product you're selling today, we're not buying.

Does that sound cynical?

We are the opposite of cynical! Nothing works for every student, every class, every period, every day, every teacher, every department, every school, every district… That is just a fact. There is no one way. But this is what makes our job interesting! We need to be eclectic, and select "what appears to be best in various doctrines, methods, or styles".

Thus, when we share ideas in this book, we are not saying they are sure to work for you just as they are. *Au contraire*! Take from those what you want, and adapt it to your own classes, your own personality, your own math background, your own school schedule, and your own beliefs. Too often, we are pressured to adopt one or another article of faith. Doing so rigidly is never a good idea. Our advice: *trust your intuition, avoid dogma, be flexible, be kind*. Of course, your intuition improves as you get more experience, but even if your intuition is "wrong", at least it is yours. Your students deserve to get the real you, not a poor imitation. If you are true to all four of these admonitions, you are sure to get better at this job.

## Setting Goals

We end with some thoughts about strategy.

### Curricular Versus Pedagogical Change

Of course, it is difficult to separate curriculum from pedagogy: switching to a problem-rich curriculum requires pedagogical changes; better pedagogy is largely irrelevant if the curricular goal is the memorizing of procedures. But if we had to choose, we would say that pedagogy is the primary frontier. Moving math education in the directions we suggested in this book will yield some immediate benefits in engagement and buy-in, even if improved pedagogy is deployed in the service of an obsolete or poorly thought-out curriculum. The resulting environment would then be ripe for curricular updates.

### Short Versus Long Term

In addition to specificity and clarity in goal-setting, it is essential to distinguish short-term from long-term goals. For example, a teacher may decide to start using number talks. A short-term goal would be to understand how this routine works, and to try it out when teaching a particular topic. A long-term goal would be to implement that sort of discourse across many topics. A department may decide to incorporate electronic graphers into their program. A short-term goal would be for members of the department to learn the basics of the software, and to teach a particular lesson with it. A long-term goal would be to figure out the implications of this tool across the whole program, and to start thinking about more powerful computer algebra systems. Over time, narrowly-focused achievements lay the necessary groundwork for broader and deeper change.

### Fast Versus Slow

When setting goals, whether for an individual teacher, a department, a school, or a district, it is most effective to be specific. What are the next incremental steps? It is not realistic to aim for simultaneous improvement on all fronts. A teacher may choose to learn a new technological tool. A department may zero in on a particular course or even particular units within a course. A school may reconsider approaches to assessment. Once the initial steps are taken, it is easier to keep moving forward. In contrast, trying to do too much too fast often backfires: when rushing, changes may not work out so well, and one may end up reverting to the status quo. Fast is slow and slow is fast.

Our experience is that step-by-step changes do pay off. The benefits are immediate: each step helps improve students' learning; it helps the teacher's professional growth; and it makes the job more interesting! We hope this book has provided you with some ideas that will help your forward motion.

## Discussion Questions

1. What approaches have you pursued to get useful student feedback? Can you think of ways to make them more effective? Think about different timing, different questions, or even different modes (e.g., one-on-one conversations.)
2. What teacher collaboration structures does your school or department expect? What structures could you add to that? (Teachers working together on different sections of the same course? Department meetings as PLC's? "Vertical alignment" of course content? Mutual classroom visits?)
3. What have you learned from conferences? From the Internet? From journals? Choose one thing you've learned to share with your colleagues.

## Notes

1. P. Liljedahl. The affordances of using visibly random groups in a mathematics classroom. In *Transforming Mathematics Instruction: Multiple approaches and Practices*, 127–144. Springer, New York, 06 2014.
2. J. Stigler and J. Hiebert. *The Teaching Gap*. Free Press, New York, 1999.
3. mathcircles.org/about/
4. Some such problems can be found at mathed.page/teachers
5. T. J. Zager. *Becoming the Math Teacher You Wish You'd Had*. Routledge, New York, 2017.
6. See Henri's article on teaching in the long period: mathed.page/teaching/long.html

# Appendix

## Special Note to Young Teachers

Dear young teacher,

The first thing we need to say is how much it lifts the spirits of your older colleagues to have you join our profession. To us, you represent the future, and we can't wait to see what you bring to your students, and to all of us. Unfortunately, as you enter this system, you are the victim of a basic flaw in its design: from day one, you are expected to do the same job as those of us with ten, twenty, or thirty years of experience. No wonder you're swamped! This job never gets easy, but it does get a lot easier than it is in the first year or two. You will never have to struggle as much as you're struggling now. Your main goal is to survive your first year or two in the classroom.

### Manage Your Time

As you no doubt have noticed, being a teacher requires you to work well past the end of the school day. That is inevitable. But you need to stay healthy and develop habits that make the job sustainable in the long run. Here are some suggestions:

- Organize the week. If you take some time on the weekend to draft a plan for all five days, your evenings will be much more manageable, as all you will have to do is tweak the plan you sketched, based on what actually happens in the classroom. You will also save time because planning all five days forces you to spend less time on each day. All this will add up to less stress overall.
- Choose whether to do that work on Saturday or Sunday. Do no school work at all on the other day, ever.
- Schedule a sufficient amount of sleep and exercise. Don't go to work if you're sick. Your students need you to get better.

Naturally, these are just guidelines. The underlying principle is that you are just as important as your students, and you need to take care of yourself.

## Grading and Standardized Tests

Do not let grading destroy you. Here are some things to think about, within what your school allows:

- Not everything needs to be graded!
- If you must grade something, be efficient.

For example:

- You might assess homework with a quick look.
- On quizzes and tests, you could just circle mistakes, without writing anything. Find ways to support individual students in learning from their mistakes, and adjust your lesson plans and future assessments to address common misconceptions.

As for standardized tests: stressing about those will tend to undermine actual learning. If you must engage in test prep, do not let it take over everything. There is no way you or any single teacher can compensate for social inequities or for students' past experiences. As much as possible, stick to your beliefs and values.

## Be Yourself

You are who you are. That is a crucial gift to your students. Let your enthusiasm for your discipline show, at the risk of seeming nerdy. Let your values show, without attempting to indoctrinate your students. Outside of class, you can share your passions: sports, movies, and music, but do not talk about your social or home life with students.

You are who you are. You do not know everything. No teacher does. It's OK to say "I don't know. I'll look into it". It is even a good thing to be wrong, sometimes, and admit it. "Oops. I was wrong. It happens". Your classroom should be a place where anyone can be wrong, teacher or student – it is a necessary stage on the path to being right!

You will learn a lot in the coming years, about your discipline, about teaching, and about how kids learn. That is what makes the job interesting over the long run. But it cannot be rushed.

## Be an Adult

Inclusivity means drawing out students who are reluctant to engage. To do this you have to be a good listener. You have to meet each student where they are. For some teachers this involves teasing, for some, clowning, for some earnestness and eye contact. For all, it involves signaling that you are their advocate, that if they will make an effort, you have their back and you will make sure they are treated with respect and protected from intellectual ridicule.

Let these qualities manifest in whichever forms are natural for you. Be aware, though, that the less the age gap between you and them, the greater the chance that some students will respond socially as if to a classmate, rather than to you as a teacher. If the students are signaling that they think you're cool, and talk to you more like a peer than a mentor, don't get fooled into thinking this is a good thing. Yes, if you're doing things right, your students will want to please you. But you should give enough cues that also signal authority. Wardrobe, body language, diction, and facial expression can all be enlisted in the cause of authority – which ones are best depends on your school culture and your personality.

You should also come well prepared with standards of classroom behavior which you will consistently enforce. Choose wisely because mere warnings and empty threats will work against you. The better your grip on the classroom and its relation to you, the more you can be zany, silly, funny, surprising, or charming in the service of mathematics teaching.

**Focus!**

When a student or a group goes off task, instead of berating them, it works best to just remind them to "focus!". The same advice applies to teachers: we are pulled in many directions. We can best address the challenges we face if we focus and prioritize. This is even more true for a beginning teacher, but it applies to all of us. Here are some suggestions:

- Set attainable goals, and forgive yourself if you don't reach them.
- If at all possible, concentrate on improving one course at a time.
- Add one teaching technique at a time to your repertoire.
- In the first year or two, do not volunteer for more tasks beyond teaching your classes. If possible, say no when asked to do such things.

There will be plenty of time to continue your growth in future years. Every once in a while, think about why you got into this line of work. Remind yourself that you will get to start over with new students next term, or next year, and every year. Everything you learn now will help all your future students, and you are learning more than you realize. If you have the energy for that, make some notes about today's classes: what worked, and what did not. Those will be handy next time you teach that material. If you don't have the energy for that, don't worry, you can start doing it next year.

### Get Help

Finally, in spite of the fact that every day you stand alone in front of your students, remember that collaboration with other adults is what will help you grow. Visit other teachers' classes if you can arrange that. You will get lots of ideas that way. Seek out buddies (other young teachers) and mentors (any experienced teachers willing to help you.) Talk to them frequently: they know the school, they know the culture, they know the administrators – and they are there every day. They are best placed to support you. They can help you prioritize: what are the big ideas in this unit? How does it connect to what the students have seen before? To what they will see later?

As time passes, you can expand your horizons and learn from books, conferences, professional development workshops, and the internet: you have joined not just a school, but a profession.

### You Are Important!

Some teachers are reluctant to miss school, even when they are sick, or when they have an opportunity to attend a workshop or conference. They feel that even missing a day or two would be a betrayal of their students. To them, we say again: You are as important as your students. You are getting healthy is, of course, in your students' best interest: it will allow you to be at your best when you return. Valuing your own professional development may seem selfish, but if you have a chance to learn something useful to your teaching, keep in mind that it will help not only your current students, but the students in your future – many, many more than are in your classes right now. And really, are you so great that missing you for a couple of days is going to permanently damage your students? Didn't you manage to survive a number of less-than-perfect subs when you were their age? It is absolutely worth it to miss school in order to get the short- and long-term benefits of your own health and professional growth.

**We are so happy you are joining us!**

# Index

abstraction 3, 27, 78, 114, 174, 184
active listening 66
administrators 10, 14, 16, 168, 187, 189, 204, 205, 212
aha! 29, 35, 45
Algebra 1 18, 32n7, 87, 119, 122, 174, 183; high failure rate 94
Algebra 2 32n7, 95, 120, 183
algorithms 13, 14, 29, 45, 100, 117–119, 132, 135, 137; long division 40
anchor *see* curricular unit
approaches, multiple 21, 76, 93, 95, 97, 207
approximation 92, 113, 119
Arcavi, Abraham 17, 17n6
architecture 29
argumentation 3, 27, 40, 74
arithmetic, mental 17, 50, 53, 58, 118–119, 178
articulation 3, 27, 54, 64, 141, 142, 150, 161
attrition 14
average: in baseball 149–150; of numbers 159, 185; speed 68, 73

back to basics 2
backlash 10
Ball, Deborah 79
Base ten blocks 79, 94, 100, 101, 113, 118, 178
Bennett, Dan xiii
bicycle riding 9
Black Pine Circle Day School xiii; Curriculum Institute 98
body language *see* language, body
boring: activities 125, 148; drills 38, 172; nature of math class 79; practice 32, 35, 118; review 172–174, 188
buddy system 4, 162
Burns, Marilyn xiii

Cabana, Carlos 180, 192n2; and complex instruction 65n2
Caffrey, Liz xiii
Cal State University, Sacramento 40
calculators 40, 92, 121–122, 138, 191; in classrooms 53, 77, 118–119, 150, 183–184; and number sense 18; and slide rule 117–118
Cangelosi, Amanda xiii
CardIAC 136
Cardone, Tina 16n5
CAS (computer algebra system) 130, 191, 206
Chakerian, Don 32, 48, 63
*Changing Minds see* diSessa, Andrea
chip trading *see* exploding dots
Chou, Rachel xiii
circulating, instructor 53, 66, 148; and doing the rounds 67
classroom heterogeneity 4, 24, 45, 167, 169, 188; within a group 70–71
classroom management 2, 50, 52, 79
classroom policies 119, 170, 174, 175, 191, 194, 195, 198
clickers 55, 159
CODAP 82
coffee spilling as strategy 38
cognitive benefits 77–79; and demands 80
Cohen, Elizabeth 65n2
cold calling 161
collaboration among teachers 141, 199–212
college: admissions 190; environment 1, 14, 16; level math 135; preparatory curriculum 32; teaching 159, 193
*College Preparatory Math* series 132
common core 2, 3, 77
communication skill 3, 14
competitions, mathematical *see* math competitions

complex instruction 65n2, 204
compliance 13, 14, 50
concentration 3, 29, 54
contemplate then calculate *see* routines
content, mathematical 17, 34, 36, 58, 117
contraries, embracing 2, 9, 11, 20, 21
conventions, mathematical 9, 20
cooperative learning 63–76, 148
counterexample 46
counterfactuals 3
COVID-19 55
Crabill, Calvin 32, 48, 63
critical thinking 10, 63; and listening 152
criticism in discussions 67, 141, 149, 150
crosswords 44, 48
cube: of a binomial 102; interlocking 113; of a number 17; root 122
Cuisenaire rods 93, 99, 100n1, 118; trains of 101
culture 10, 145, 211–212; classroom 13, 53, 72, 154, 172, 204; departmental 51, 69, 190, 196–197; mathematical 9; wars 2
Cuoco, Al 5n2
curricular unit 183; sequencing within 183–185; starting with anchor 184
curriculum: framework for 180–181; mapping 181–183; pruning 177–178; *see also* problem-based, sequencing, spiraling

dead ends 72–75; flaming 72
debate: absurd 9; on calculators 53, 122; classroom 103, 203; hostile 2; on learning styles 78; polarized 53
decimals 30, 54, 100, 151; repeating 40
definitions: comprehension 29; learning to use 68, 73; supplying 67; starting with 184
*Designing Groupwork* 65n4
Desmos 45, 81, 97, 119, 138, 196; Activity Builder 45; Marbleslides 126–127
dichotomy, false 162, 175
direct instruction 11–12, 20, 33
discovery method 2, 20, 31, 142–143, 183; activity 12, 24, 32, 35; versus direct instruction 11–12, 33

discussion, Socratic *see* Socratic discussion
diSessa, Andrea 138n14
dissections, geometric 24, 27, 29
distributive law, property or rule 14–16, 53, 102, 119–120, 180, 185; and factoring 15, 102, 130, 178
district *see* school district
divisibility 34–35
division 2, 177–178; of fractions 40; long 40, 117–120; polynomial 102
dogma 162, 205–206
drill, non-random *see* non-random drill
Dudeney, Henry 105, 116

eclecticism 204–205 *see also* approaches, multiple
educational policy 1
eggs, mistakenly put in one pedagogical basket 58
Elbow, Peter 9n2, 10, 11
elementary school 1, 113, 142, 201
encouraging 10–11, 152–153, 158, 160
engagement, intellectual 4, 19, 21, 50–53, 137, 142, 162
Erdös-Strauss conjecture 31n2–3
errors: computational 168, 191; intentional 41, 152, 155; logical 74; tracing 68, 74
estimation 17, 55, 118–119, 122, 178
evening star 9n4
Exeter *see* Philip Exeter Academy
experimentation 28, 134–135, 185
exploding dots 82, 93
exploration 1, 20, 24, 32, 34, 53, 63, 101, 134–137
exponential growth 40, 180
extension of a problem 39, 47, 100, 102, 104
extremes of math education 2, 11, 38

factor 15, 17, 81, 102, 114, 130, 157, 177, 178; greatest common 38
fads, pedagogical 10, 203–205
Farrand, Scott xiii, 40, 52, 53, 155
Ferdinand, Vic xiii
Fibonacci numbers 25n1, 80
*Five Practices* 75, 147n3, 149, 157
FOIL 16

fractions 83, 94, 98, 106, 185; Egyptian 30, 48; equivalent 15; multiplying 84; simplified 13, 102; well-chosen rectangle 81
free riders 65, 70
Freeman, Gregory 9n1
frustration 43–45, 63, 118, 167
function diagrams 81, 89–90, 94

games 57–58, 95n16, 105; electronic 78–79
Gardner, Martin 42, 48, 105
geoboard 32n10, 81–82, 96, 110–112, 115, 184; circular 112; virtual 132
GeoGebra 82, 96n18, 119, 128, 130, 132n6, 133, 196
*Geometry: A Guided Inquiry* 32–33, 63, 185
glue, conceptual 1, 138
Goldenberg, E. Paul 3n2
Golomb, Solomon 105, 116
grading 190, 194–196, 204; guidelines 190–192; and oral exams 195; and test corrections 192–195; and time management 210
graphical solutions 119–122, 149, 181
grid paper 27–28, 35, 83, 96
group work dynamics 67, 69, 75, 148–150; intervention 68, 72–75, 172
group-worthy activity 34, 36
groups, theory of finite 142
groups, visibly randomized 65–66, 76, 199n1; frequently change 66
growth: exponential *see* exponential growth; intellectual 23, 75; mindset 203–204; professional 82, 201, 207, 211–212
guess and check *see* trial and error
guided inquiry 32–34, 48, 63, 185
guidelines: lesson planning 181, 183, 184; philosophical 167; from Project SEED 143; for teachers 4, 46, 51, 67; time management 209

habits 47, 50, 56, 135, 192, 197; of mind 3–5, 23, 56n3, 114; of teachers 187, 209
hands-on activities 12, 37, 67, 99, 114–115, 132, 200

hearing: answers without questions 39; from everyone 143; peripheral 66; questions 51
Helft, Shira 80, 87
heterogeneity *see* classroom heterogeneity
Hiebert, James 201n2
hints 12, 17, 36, 38, 43–46, 53, 71, 75, 143, 157, 159, 195
Holt, John 152
homework 37, 65, 76, 146, 161, 170–177, 187, 194–197, 201, 210; going over 49, 52, 64, 71, 160, 172, 200; lagging 170–171, 174–177, 186, 204; and test corrections 175, 197, 200
honors classes 4

ideologies 2, 12
*Illustrative Mathematics* 32, 55n1
inclusiveness 151, 161, 210
inequalities: educational 4, 120; mathematical 150
informal: classroom 50–51, 58, 67; reasoning 22, 29, 183; versus formal 12, 50, 58
initiative 3, 23, 77, 80, 82, 142
inquiry-based learning 31, 72; guided *see* guided inquiry
institutionalization 18–21, 149, 151
instruction, direct *see* direct instruction
internet: community 203, 212; instruction 79, 204; searching and sharing 37, 174, 205, 207
intuition 30, 67, 73–74, 180; developing 18, 45, 128; geometric 130; teacher's 206

Johnson, David 145–146
Johntz, William 142
justification, mathematical 3, 158, 181, 193

Kandl, Frances xiii
Kelemanik, Grace 56n3
kinesthetic activities 56, 57n4, 83n8, 129, 184
Krall, Geoff 38

Lab gear 100–103, 178, 191
language: body 69, 151, 152, 155, 211; mathematical 3, 10, 20, 55n1; programming 27, 110, 135, 136; word problem 41
Lautze, Richard xiii
learning tools 46, 77–98, 132, 137, 173, 202
Lee, Meghan xiii
Leinwand, Steve xiii
lesson plan 12, 50, 96, 147, 148, 182, 187, 201, 210
Lester, Tom 156
Liljedahl, Peter 65, 172, 199
linear: equation 15, 119, 131; function 18, 29, 39, 183; growth 40, 86; systems 122–125, 150
locus 56–57, 128–129
Loewinsohn, Briana xiii
Logan, Rachel 65n2
Logo (programming language) 27, 110, 135
Lucenta, Amy 56n3
Lucius, Lisa 9n1

Ma, Liping 48
magical hopes 79, 97
Marbleslides *see* Desmos, Marbleslides
Mark, June 5n2
math competitions 37, 85, 168, 175
math education research *see* research, math education
math teacher circle 202–203
math wars xi, 2, 3, 9, 16, 48
*Mathemania* 85n10
*Mathematical Carnival* 42, 48
*Mathematical Language Routines* 55n1
mathematicians 21, 32, 85, 94, 142, 168, 169, 203; mathematician's math 2
*Mathematics for Equity* 65n2
*Mathematics Teacher* journal 9
McCallum, Bill 55n1, 182; estimation 55
memorization 9, 12–15, 35, 36, 45, 49, 52, 56, 100, 114, 162, 181, 204, 206
mental arithmetic *see* arithmetic, mental
message, implicit 36, 56, 66, 154, 172, 187, 191

meta-curriculum 3, 4, 23, 29, 34, 36, 56, 114, 156, 174
Meyer, Dan 38
*Mindstorms* 78n1
Moon, Susan xiii
morale 11, 29, 68
morning star 9n4
multiple approaches *see* approaches, multiple
multiplication 41, 94, 178; algorithm 177, 185; facts 118; of fractions 84; rectangle model 15, 17, 80–81, 102, 130; relation to number sense 18
multistep 38, 127

National Council of Teachers of Mathematics *see* NCTM
NCTM 2–4, 48, 75, 76, 158, 163, 202
Nelson, Scott xiii
new math 2, 194, 204
Newton, Isaac 21
no child left behind 2
non-random drills 41–42, 47
norms, behavioral 51, 58
notation 13, 14, 20, 183, 184
note taking 18–19, 162; teacher controlled 21; by teachers 148, 156, 187, 192, 202, 211
number lines 83, 89, 93–94, 98
number sense *see* sense, number
number talks *see* routines

Ohio State University xiii
opposites 11–13
ownership 3, 23, 34, 77, 142

pace: change of 46, 50, 58; of course 11, 31–32, 66, 70, 167, 169, 172, 197
pandemic *see* COVID-19
Papert, Seymour 78, 103, 110, 135
paradox 24, 25n1, 29, 42
parallelogram 26–27, 35, 104
partial solutions 34, 43, 45
passivity, combating 53, 56, 64
paths, curricular 2
pathways, multiple 34, 43–45, 54, 69, 130
patronizing 70, 152
pattern blocks 82, 106–107; virtual 132–133

Pemantle, Oscar xiii
pencil and paper computation 53, 77, 79, 118–121, 132, 177
percentages 40, 72–73, 134
perimeter 27–29, 103, 106, 151
perpendicular 57; construction of 130; bisector 128, 129
perseverance 29, 44, 46
Pershan, Michael xiii
Phelps, Steve 96n18
Philip Exeter Academy 32
Photomath 130
physical: aspects of teaching 151, 160; challenge 127; learners 78; materials 78, 96, 99, 105, 114, 115, 128; motion for students 79; probability experiments 135; versus virtual materials 132–133, 138
piano scales *see* scales, piano
Pickford, Avery 3n2
PLC *see* professional learning community
polarization 2, 53
policy: classroom *see* classroom policies; educational *see* educational policy
poll the class *see* routines
polyomino 27–28, 105, 116
precision 3, 136, 156
preservice teachers 4, 40
*Principles to Actions* 75–76, 149, 158
problem solving 3–4, 20–47, 52, 63, 114, 137, 147, 179–180, 184, 204; techniques 46
problems 19, 23–48, 49, 74, 9–93, 102, 132, 168, 177–178, 181, 191, 194; anchor 185; -based curriculum 29–30, 46, 63; challenging 34, 53, 66, 137; creating 37, 195, 202, 297; head 54–55; interesting 18, 32, 52, 131; low-threshold high-ceiling 27, 29, 34, 46–47, 73, 136; open-ended 38; and puzzles 42, 116; rich 24, 34; sequencing 33; well-chosen 13; word 40–41, 119–120, 132
procedures 2, 64, 79, 100, 129–130, 191; applying 23; memorizing 49; obsolete 177

professional development 4, 9, 19, 45, 53, 201, 212
professional learning community 201–202, 207
Project SEED 142–146, 157–160; response techniques 143–144, 160; URL for materials 143n1
proportion 15, 153
Propp, James xiii, 55, 158
puzzles 24–27, 42–47, 57, 82, 96, 101–106, 115–116, 126
Pythagorean theorem 15, 29, 42, 96, 98, 110, 112, 180, 183

quadratic: equation 119; function 126, 183; relation 29
questioning 23, 70, 86, 146, 157

reasoning 27, 29, 72, 74, 157, 178, 180
reform 2, 9, 10, 97
representations: multiple 15, 52, 82, 93–97, 169, 173, 180, 197; visual 15, 95–97
research, math education 10, 65, 76, 199, 203–204; overgeneralized 10
response time 156
reverse process 15, 37, 89, 181
review, eternal 11, 169, 186, 191, 198; and constant forward motion 169; and cumulative tests 191; and spiraling 186
rich activities 25, 32–35, 45, 47, 169
routines 41, 55–56, 58, 119, 206; contemplate then calculate 56; number talks 54; openers 52; poll the class 55; think first 52; think-pair-share 55

Safa, Parisa xiii
scales, piano 16
schema 3
school district 10, 120, 168, 187, 205, 206
Seeger, Pete 12
sense, making 17, 36, 40, 41, 95, 102, 112, 118, 158, 178, 181; number 17, 18, 118, 121; operation 18; symbol 18

sequencing: curriculum 167, 178, 182–186; discussion 147–149, 153; lesson 39, 95
Serra, Michael 128n6
shapes, geometric 24, 27, 103, 105
shortcuts 14, 16, 17, 42, 130
silence 64, 155; practicing in 1, 13, 35, 49; teaching in 160
skills, basic 2, 6, 137
slide rules 77, 117, 121
slides, lecture 19–20
slides, simulated on Desmos 45, 126–127
slope activity, opener 52
Smith, Margaret 76, 147–148, 163
Snap! 82, 135
Socratic discussion 141–142, 145, 159, 161
Solomon, King 12
spiral curriculum 53, 185–186
spreadsheets 77, 80, 133–135, 138, 150
square root 30, 91, 112; computation of 117, 121
staging 151
standards 183; Common Core 2–3, 77; NCTM 2–3; based grading 190, 204
Stein, Mary Kay 76, 147–148, 163
Stein, Sherman 32, 48, 63
Stigler, James 201n2
Stony Brook University 10
story tables 81, 87–89
straightedge: and compass 127–129; and ten-centimeter circle 92
stuck, students being 47, 63, 67–78, 148
student centered: activities 34, 36; classroom 79, 168
Szydlik, Jennifer xiii

tangrams 103–106
Tanton, James 83, 158
teacher preparation programs 9
teaching, eclectic 10, 21, 204–205
*The Teaching Gap* 201
ten-centimeter circle *see* straightedge
terminology 13, 20
test corrections 175, 191–197, 200, 204
testing 2, 181, 191; standardized 2, 37, 195, 197, 198, 210

think-pair-share *see* routines
three-act problem 38, 48
time pressure 37, 168, 170n1, 190, 193–196
tools *see* learning tools
tracking 2, 4, 30
transition: concrete to abstract 114n7, direct to open ended 33, 37; informal to formal 51; mental to paper 114
trapezoid 35–36, 39, 104
trial and error 15, 46, 69, 88, 117, 125, 130, 149–150, 157
triangles 27, 35, 39, 81, 86, 104, 109–113; isosceles 24, 105; right 25, 94, 105; similar 26
tricks: mnemonic 16–17, 22 *see also* shortcuts; pedagogical 144
trigonometry 32n7, 93–94, 130, 183, 186; tables 121
turn and talk *see* routines, think-pair-share
turtle: geometry 80, 134, 135n1; graphics 110, 135; quadratic 126

University of Washington's Whiteley Center xiii
University of Wisconsin xiii
Urban School of San Francisco xiii

variables, mathematical 3, 14, 35, 46, 69, 75, 94, 150, 159, 184; environmental 187, 205
Venn diagrams 73, 77, 84–86, 97
visits from colleagues 13, 75, 141–142, 162, 200, 207, 212

Wah, Anita xiii, 48, 198
warm-ups 52–55, 171
Weltman, Anna xiii
whiteboards 55, 65, 148, 204
Whiteley Center *see* University of Washington's Whiteley Center
worksheets 22, 35, 98, 125, 132, 157
wrap-up 20, 21, 37, 153
wrong answers 74, 86, 149–154

Zager, Tracy 203

For Product Safety Concerns and Information please contact our EU
representative  GPSR@taylorandfrancis.com
Taylor & Francis Verlag GmbH, Kaufingerstraße 24, 80331 München, Germany

www.ingramcontent.com/pod-product-compliance
Lightning Source LLC
Chambersburg PA
CBHW060511300426

44112CB00017B/2623